D1554703

Wild Plants of the Pueblo Province

Wild Plants of the Pueblo Province

EXPLORING ANCIENT AND ENDURING USES

by William W. Dunmire and Gail D. Tierney

Foreword by Gary Paul Nabhan

Musem of New Mexico Press
Santa Fe

Line-art Illustrations by Gail D. Tierney
Color photography by William W. Dunmire

Copyright ©1995 Museum of New Mexico Press. Photographs by William W. Dunmire copyright the photographer. *All rights reserved.* No part of this book may be reproduced in any form or by any means whatsoever without the expressed written consent of the publisher, with the exception of brief excerpts embodied in critical reviews.

Manufactured in Korea.
Project editor: Mary Wachs
Design and production: David Skolkin
Cartography: Deborah Reade

Library of Congress Cataloguing-in-Publication Data Available.
ISBN 0–89013–282–8 (cloth); 0–89013–272–0.

Museum of New Mexico Press
Post Office Box 2087
Santa Fe, New Mexico 87504

10 9 8 7 6 5 4 3 2 1

Cover Photograph: Globe-Mallow
Frontispiece: Cane Cholla at Bandelier National Mounument

Contents

Foreword vii
by Gary Paul Nabhan
Preface xi

PROLOGUE
1

CHAPTER 1
A Region of Enormous Natural Diversity 9

CHAPTER 2
The Ancestral Puebloans 13

CHAPTER 3
Contact, Colonization, and Catastrophe 37

CHAPTER 4
Modern Descendants 45

CHAPTER 5
Indicator Plants as Living Artifacts 73

CHAPTER 6
Plants and Plantcraft 83
Trees 95
Shrubs 122
Grasses and Grasslike Plants 153
Herbaceous Plants 163

CHAPTER 7
Ethnobotany in New Mexico 227

CHAPTER 8
Recent Modifications to the Landscape 241

CHAPTER 9
Other Places to Visit 249

An Annotated List of Useful Plants 254

Bibliography 267

Photography Credits 279

Index 281

About the Authors 290

FOREWORD

THE PUEBLO PROVINCE—including the tribal lands drained by the Upper Rio Grande — is perhaps the region in North America where diverse cultural relationships with plants remain most intact. The ancient ethnic traditions of northern New Mexico have not merely survived, they have somehow thrived despite the economic upheavals and environmental degradation of the last five centuries. To be sure, the degree of dependence upon indigenous plants has diminished in some pueblos more than others, but the native flora is never too far out of sight, out of mind, or out of spirit. The fragrance of "cedar" smoke, the taste of wild celery, and the sound of a gourd rattle are never too far away.

As a testament to these tenacious cultural connections to a wild and verdant world, Gail Tierney and Bill Dunmire have documented the prehistoric, historic, and contemporary uses of 300 species in the southwestern flora. Over their many years as students of the plants and peoples of New Mexico, they have diligently collected information from many far-flung sources. Now they have succeeded in forging these various elements into a well-tempered compendium full of "news" from ancient and enduring lifeways.

There is authority to this work, but it is inviting rather than intimidating. What distinguishes this book from other, more casual inventories of edible, medicinal, or dye plants is that Dunmire and Tierney have painted a backdrop for each leafy plant, its uses and its lore. They have placed each root into a deeper ecological and historical context so that its natural and cultural connections to place are made evident. Ecology is not considered extraneous material to cultural studies but an essential element of the human story for this region.

Unless one understands the ecology of each particular species, it would be difficult to determine which plants are reliable indicators of the locations of springs, seeps, or high water tables. Dunmire's work as an ecologist has given him a clear picture of how each species fits into the ecozones, or biotic communities, associated with the semiarid uplands of the Southwest. Unless one knows the region's prehistoric subsistence technologies, and how historic contact with Athabaskans, Hispanics, and Anglos affected them, it would be difficult to determine whether a particular plant-processing technique is ancient or introduced. Tierney's collaborations with archaeologists as well as contemporary healers and craftsmen has provided her with insights which few other ethnobotanists have. Together they have woven a fabric of ecological warp and anthropological weft as durable as any Pueblo or Navajo rug.

Reading this work, I could imagine how, over the centuries, individuals from different cultures enriched each other's knowledge of valuable plants found in their homelands. I could hear them talking to one another around campfires, during healing rites, or out in their fields. Few of these plants have escaped any use by Hispanic and Anglo newcomers to the region, and fewer still are known by only a single Native American culture. In short, they are truly part of a multicultural heritage of knowledge about the natural world and need to be conserved as such.

At the same time, the authors have been careful not to divulge ceremonial or otherwise sensitive information about a value-laden plant-use unique to a single pueblo or clan. Again, they show their respect for the context of traditional knowledge instead of treating every empirical observation about native plants as an "information bit" suitable for driving out onto the information superhighway.

We now know that historic Puebloan peoples were not mere passive collectors of wild plants; they also transported, transplanted, pruned, fire-managed, and selected certain useful plants much as horticulturists, foresters, and plant breeders do today. Some southwestern plants long believed to be part of the natural vegetation are now recognized as incipient domesticates; that is, their seeds or shoots may have been carried from one pueblo to the next, where they were planted, hybridized, or protected for generations.

Try to fathom where the Zuni obtained the progenitor of the domesticated husk tomatillo, *Physalis philadelphica*, which has not been collected elsewhere in northern New Mexico. Or imagine prehistoric terraces scattered across the Southwest that were planted with a tequila-like agave, selected for its food and fiber qualities. These plants, along with turkeys and macaws, have stories hidden within them, stories of long-distance trade, ritualized tending, and celebration.

This book unearths many such stories, but perhaps not all of them. Rather than being a last word on Pueblo ethnobotany, it will stand for years as the most useful point of departure for new explorations. I can imagine a Native American linguist reading it and becoming inspired to learn the origins of names (and their meanings) for a particular set of plants. I can also hear a Pueblo storyteller incorporating more and more plant lore into the parable he or she tells a group of children gathered around a fireplace in wintertime. Gail Tierney and Bill Dunmire both honor the elder nature-literate storytellers and offer a complementary body of knowledge to younger generations who can now use this book as one more touchstone for maintaining their people's connections with plants. And that is as it should be.

Gary Paul Nabhan
Director, Native Seeds/SEARCH
Tucson, Arizona

PREFACE

THIS BOOK ORIGINALLY was conceived by the authors as a guide to some of the common trailside plants of Bandelier National Monument and the Pajarito Plateau and their prehistoric and recent uses by Indians living in this area. As we discussed the project with others, it soon became apparent that a larger story begged telling—a chronicle of plant uses that encompassed all of the traditional territory of the nineteen modern pueblos of New Mexico, centering in the middle Rio Grande Valley.

We were encouraged by a number of Puebloans who recognized the educational value of this project and who shared some of their personal knowledge of the wild plants or allowed us to visit their old-style gardens. Among them we are grateful to Felipe Lauriano of Sandia Pueblo, Lorrain Loncasion of Zuni, Bill Martin of Cochiti, Ramos Oyenque of San Juan, Peter Pino of Zia, and Pat Toya of Jemez. Other Puebloans who gave us helpful suggestions were Gregory Cajete, James Hena, Robert Mora, Joe Sando, and William Weahkee. We are especially grateful to the National Park Service Southwest Regional Office ad hoc cultural affairs advisory committee consisting of Ed Natay, Gary Roybal and Virginia Salazar. They reviewed much of the manuscript and advised us on matters of cultural sensitivity.

Others who provided critical review of portions of the manuscript were Craig Allen, Bandelier National Monument; Brett Bakker, Native Seeds/SEARCH; Richard Ford, University of Michigan; David Hafner, New Mexico Museum of Natural History; Glen Kaye, National Park Service; Loren Potter, Professor Emeritus, University of New Mexico; Douglas Schwartz, School of American Research; and David Stuart and Wirt Wills, both of the University of New Mexico. We are indebted to all reviewers with the caveat that none bear any responsibility for the authors' errors or omissions that surely will creep into print.

Staffs at the four parks we regularly visited to document trailside plant examples invariably went out of their way to assist us. Individuals deserving special mention include Roy Weaver, Craig Allen and Chris Judson at Bandelier National Monument; Paul Henderson at Petroglyph National Monument; Nathan Stone at Coronado State Monument; and Frank Gachupin at Jemez State Monument.

We particularly wish to thank the museums and other institutions that aided us. The University of New Mexico Herbarium loaned numerous plant specimens that were used as models for Gail's line drawings; Assistant Curator Jane Mygatt helped in many ways. Others at UNM who furthered our cause were Professors J. J. Brody, William Hadley, Rich Hollaway and Carl White, along with Stanley Smith at New Mexico State University. Historical photograph and research assistance came from the Laboratory of Anthropology through Laura Holt, Patricia Nietfeld, Willow Powers and Louise Stiver; from the National Park Service Southwest Regional Office through Adeline Ferguson, Glen Kaye, Janet Orcutt and Robert Powers; from Chaco Culture National Historical Park through Dabney Ford and Philip LoPiccolo; from the Maxwell Museum of Anthropology through George and Gloria Duck and Natalie Pattison; from the School of American Research through

Katrina Lasko; from the Office of Cultural Affairs, Historic Preservation, through Louanna Haecker; and from the Museum of New Mexico Photo Archives through Richard Rudisill. Additional help came from the Zuni Archeological Program through Carol Brandt, Zuni High School through Patricia Allen, and the Ellis Archives through Andrea Ellis.

Special thanks go to Blair Clark for his photographs of specimens from the Laboratory of Anthropology collection and to Gary Nabhan, who contributed his insightful foreword.

Credit for polishing the manuscript, creating an elegant book design, and all else required to transform words and pictures into a presentable book goes to the ever-cheerful staff of the Museum of New Mexico Press, especially to Editorial Director Mary Wachs and Art Director David Skolkin.

This book could not have been researched and written without the constant encouragement and support of our spouses, Martin Tierney and Vangie Dunmire. These good, loved, and loving people suffered through more than occasional bouts of grouchiness or panic, never complaining about our absorption by the work at hand. Thank you, Vangie; thank you, Martin.

PROLOGUE

*U*P ON THE MIDDLE SLOPES of the Jemez Mountains in New Mexico, the Pajarito Plateau is shaped by a series of gently tilting mesas dissected by steep-walled canyons—some watered year-round, others dry except during flash floods. This land, clothed in richly diverse vegetation and nearly as bountiful today as it was down through the ages, has long been a home to American Indians, especially during the past eight hundred years. Indeed, there may be more prehistoric ruins and other archaeological sites concentrated here per square mile than in any comparable place in the Southwest.

To the east and south the land slopes down to the Rio Grande, which courses through a series of broad floodplain valleys interspersed with occasional boxlike canyons. Whereas the Pajarito and the Jemez Mountains above retain a wilderness flavor even now, the character of the valley below has largely been transformed by human habitations. In fact, the great majority of modern-day New Mexicans live within the confines of the Rio Grande Valley and its imme-

In time there was a movement down from the plateau country, with its abundance of life-sustaining wild plants and animals, into the valley of the Rio Grande and its major tributaries, where water always was available for crops and where communal villages, later known as *pueblos* (the Spanish word for "towns"), would become established. But wild plants remained ingrained in the fabric of Puebloan society, and later, when the Spaniards began to move up the Rio Grande corridor, Pueblo people shared their knowledge about some of the plants. Many of these uses were adopted by the newcomers and helped them to prosper. Today, people from the modern cultures that thrive along the Rio Grande continue to collect certain wild plant parts and use them in a variety of ways, although the various plants and their uses may vary from pueblo to pueblo.

This book, then, is about these useful wild plants and how some of the more important species figured in the lives of the people who once lived here and in the lives of their descendants. It will aid those who visit places such as Bandelier and Petroglyph national monuments and Coronado and Jemez state monuments in identifying the most common plants found along the many paths of these parks. The central focus of the book is the centuries-old connection of these plants with humans and their daily needs.

Bandelier National Monument is near the center of and makes up about one-sixth of the 300-square-mile Pajarito Plateau. Archaeological records from Bandelier provide much material for our interpretation of the story of plants and people, but information from nearby related sites such as Jemez Cave, Puye Ruins, Otowi Ruins, and ancient Arroyo Hondo Pueblo, just five miles south of Santa Fe, also is important. And more recent historical data derived from peoples of the modern pueblos of the Rio Grande and its tributaries—especially San Juan, Santa Clara, San Ildefonso, Cochiti, and Jemez pueblos—have been incorporated wherever possible.

Although written for those who have no formal training in botany or anthropology, this book is based on material from nearly three hundred original technical reports and manuscripts, both old and recent, coupled with interviews with living authorities, both academic scholars and people of the land. Thus, we hope what we have to say will be fresh and of value to serious students as well as the lay public.

How To Use this Book

THE BOOK IS ORGANIZED so the reader can learn both why plants grow where they do on the Pajarito Plateau and middle Rio Grande Valley and about the shifts in vegetation that have taken place over the millennia, including recent human modifications to plant habitats. First, we address the profound influences that changes in climate and plant communities are thought to have had on prehistoric settlement patterns within a region covering the Rio Grande and its tributaries, from Taos Pueblo south to Isleta Pueblo and west into the Rio Puerco and San José river drainages. This region, which we refer to as the *New Mexico Pueblo Province*, incorporates several mountain ranges, among them the Jemez, Sangre de Cristo, Sandia, and Zuni mountains. Our description of the uses of wild plants by the ancestral Pueblo people is based upon archaeological evidence, which in turn is compared with recorded uses by Puebloans living in this region in historic times. Finally, accounts of individual plant species that have important cultural tie-ins and are relatively common are featured at the core of this book.

Our field investigations at Petroglyph, Coronado, and Jemez covered plants and plant communities likely to be seen from all the main public trails. At Bandelier we concentrated on the Ruins Trail, the Frey Trail, the Falls-Rio Grande Trail, the Upper Crossing Trail, and, of course, the scenic and archaeologically significant path that leads to the ancient village of Tsankawi. This book is meant to be a companion on the day walks and hikes at these public parks that

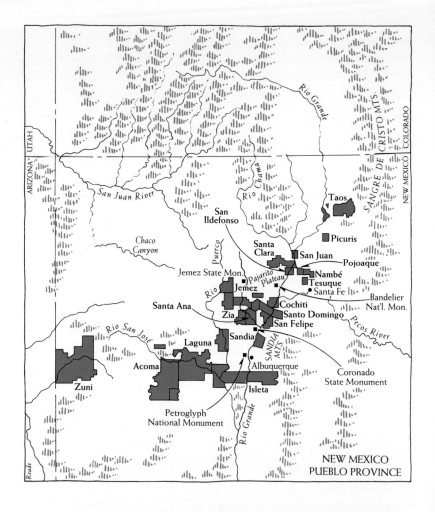

visitors and residents alike would most likely take, and it emphasizes plants easily seen at the above locations and throughout the New Mexico Pueblo Province.

More than a thousand species of wild plants have been recorded from the New Mexico Pueblo Province. Here, we focus on just over sixty that are both reasonably common and have a "people story" to tell. The species accounts beginning on page 95 are grouped in the sections on trees, shrubs, grasses, and herbaceous plants.

Imagine that you're walking along the Ruins Trail at Bandelier. As you approach Long House Ruins, you notice shrubs growing on either side of the path. They stand nearly four feet tall, have narrow gray-green leaves, and about half of the plants have peculiar seed pods with four papery wings surrounding each seed. Turn to the section on shrubs and leaf through the illustrations. No other shrub on the Pajarito Plateau resembles this plant. You'll have no trouble identifying it as fourwing saltbush. Then you'll learn that the hard wood of its branches was once employed to fashion arrowheads, and, more recently, its crushed flowers have been used by people of Zuni Pueblo to relieve the effects of ant bites. You'll also read about how fourwing saltbush is an indicator plant of formerly disturbed sites, one of the reasons it is so common in Frijoles Canyon. And if you hike to Tsankawi, you'll come across a dense stand of saltbush growing on top of the mesa, masking the ruins of some 350 rooms, most of them still buried.

PREHISTORIC TRAIL LEADING TO TSANKAWI, BANDELIER NATIONAL MONUMENT.

An annotated checklist of plant species with known uses by past and present Puebloans from within our region is included at the end of the book. Intended for the scholar, coverage in this list is much more extensive than that for the more common plants in the main text and includes for each species a summary of the various uses reported in the technical literature.

Respect for the People and the Land

WE ASK YOU TO REMIND YOURSELF that the people who make up the heart of our story are likely to have a view of the natural world different from yours. For example, most traditional Pueblo Indians hold that people, plants, animals, and spirits are all interconnected in a seemingly unbroken circle of being. Many wild plants have sacred values to modern Puebloan people. We've tried to avoid dwelling on plant uses whose mention might distress, or even offend, native cultures in this region.

As you walk the trails, remember that all natural and cultural objects—wildflowers, Indian artifacts, even rocks—should be left in place for others to enjoy; in fact, it's against the law to violate this rule in any national or state park or monument. Remember, too, that some of the places you visit may be archaeological sites to you but have spiritual importance to contemporary Puebloans.

Today, with more and more people seeking recreation in natural settings, the ethic of "take nothing but pictures, leave nothing but footprints" ought to extend beyond park boundaries. As wilderness becomes ever more scarce, we must recognize the fragility of these places, cherish what is left, and leave it intact. For this our grandchildren will thank us.

1
A REGION OF ENORMOUS NATURAL DIVERSITY·

*F*EW PLACES IN WESTERN NORTH AMERICA rival the enormous plant and animal diversity found along the middle Rio Grande, adjacent mesas, and the mountains within ten miles or so of the river. In fact, over 700 species of wild plants have been recorded from Bandelier National Monument alone. The reasons for this are straightforward. Elevations in the Rio Grande corridor range from five thousand to more than ten thousand feet; the topography contrasts sharply from flat valley bottoms and mesa tops to steep-walled canyons and slopes, and contains many different geologic substrates and soil types. These variables translate into a wide range of moisture and

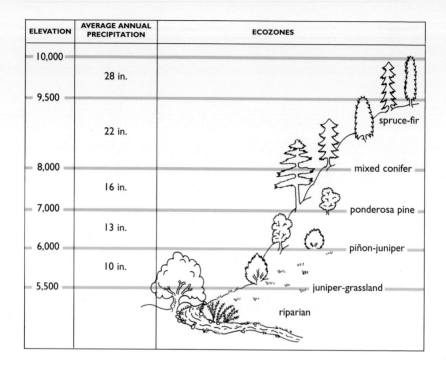

temperature patterns, and in nature such climatic variability is the key to determining what grows where.

High plant, and coincident animal, diversity and abundance may have been key in drawing the Pueblo people of the Four Corners region eastward in the twelfth and thirteenth centuries. Certainly, this accounts for population growth in the region from the time of Pueblo settlement after about A.D. 1300. In a study done near the present Cochiti Reservoir, just south of Bandelier National Monument, 218 different plant species were observed growing in several small vegetation research plots within Cochiti Canyon. Of these, ninety-six, nearly half the total, had plant parts that are potential food for humans (Tierney 1979).

Besides providing ample sources for food, high biodiversity ensured a broad selection of plants for fiber, implements, and the construction of dwellings. Finally, this diversity offered the all-important spectrum of plants whose medicinal properties formed the basis of American Indian herbal medicine, so vital for the well-being of ancestral Puebloans and widely relied upon in the present.

The vegetation just above the shrubby semidesert grasslands bordering the Rio Grande is dominated by two species of evergreen conifers—one-seed juniper and piñon pine. This piñon-juniper plant community, or *ecozone*, especially where it overlaps with the ponderosa pine ecozone at an elevation above six thousand feet, provides a habitat for more species of edible plants than any other major biological community in the region (Drager and Loose 1977) and in all New Mexico. With the skillful development of hunting, gathering, and agriculture, the piñon-juniper ecozone eventually came to comprise the core area for human habitation in the region.

Archaeology is unable to conclude with certainty whether it was this plant-animal diversity that drew the Pueblo people into the Rio Grande Valley or whether they

THE PIÑON-JUNIPER ECOZONE COMPRISES THE CORE AREA FOR PREHISTORIC HUMAN HABITATION.

migrated here for other reasons. Clearly, the story is much more complex and multifaceted. But the fact remains that the Pajarito Plateau, the Rio Grande Valley, and the adjacent watershed eventually would provide a bountiful home to evolving cultures that yet thrive in modern-day New Mexico.

Suggested Reading

Dick-Peddie, William A.
1993 *New Mexico Vegetation: Past, Present and Future.* Univ. of New Mexico Press, Albuquerque.

Kelley, Edmund N.
1980 *The Contemporary Ecology of Arroyo Hondo, New Mexico.* School of American Research Press, Santa Fe.

2
THE ANCESTRAL PUEBLOANS

From Hunters to Foragers

*T*HE EARLIEST EVIDENCE of people inhabiting the land surrounding the northern and central Rio Grande Valley comes from radiocarbon remains collected at ancient campsites dating to about 8000 B.C. Evidence from other sites in the Southwest suggests that people camped in or passed through this country at an even earlier time, perhaps much earlier, but precise archaeological data are lacking for the New Mexico Pueblo Province (also referred to as "our area").

We don't know much about the individual and social lives of these earliest recorded people, but we do know their subsistence depended largely on wild animals. The remains of mammoths, camels, tapirs, horses, an immense species of bison, and other large mammals long since gone from North America are directly associated with many archaeological sites of this era, attesting to a hunting-scavenging way of life that was no doubt supplemented by the gathering of wild plants.

Despite being far to the south of the great continental ice sheets that recently had covered much of North America, this area—its climate, vegetation, and animal life—during the early postglacial period was vastly different from our present-day environment. The glaciers of the north had mostly dissipated, but the Rio Grande Valley and adjacent plateaus still were relatively cool and wet, particularly in the summer. Some of the low-lying basins contained large lakes where today there are only beds of alkaline silt and sand. Coniferous forests thrived at elevations much lower than where they now occur. Below the Sangre de Cristo, Jemez, and Sandia mountains the land leading up from the Rio Grande would have been covered with pine and deciduous woodlands interspersed with savannas of tallgrass prairie and sagebrush. These conditions prevailed in sharp contrast to the piñon-juniper and desertlike shrubby vegetation that have characterized this area for the past four thousand years or more.

Perhaps the greatest hallmark of this time was the so-called *megafauna* that had dominated the land for many millennia. In our area the giant mammals included herding beasts such as mammoths, horses, and camels, as well as ground sloths, giant armadillos, saber-toothed cats, oversize bears, and a number of other genera.

But within a relatively short period the populations of many of these animals declined precipitously, and by 6000 B.C. most of them had virtually disappeared, leaving only elk, deer, bighorn sheep, and pronghorn and a smaller race of bison among the large browsing and grazing mammals in New Mexico. Was excessive human predation the culprit in this episode of species extinction, or did spells of severe drought within a general trend toward a drier climate, with associated changes in forage and range vegetation, cause the demise of the megafauna? Both hypotheses have their advocates today, as they've had since Darwin's time. In all likelihood, however, environmental stress was the driving force, with ever-increasing human predation delivering the coup de grace.

In any case, the effect of the megafauna extinction upon people living throughout the continent must have been profound. The early hunters would have borne the brunt of this rapidly declining supply of quarry and might even have experienced a population crash themselves before they learned to adopt new subsistence strategies. They appear, in time, to have shifted away from their decidedly nomadic life-style, which had depended upon the movements of large animals.

The new way of living allowed for the occasional establishment of transient campsites that would have been occupied on a seasonal basis. At the same time the depletion of large herds of herbivores (with the exception of the plains bison, whose numbers actually increased during this period) must have resulted in at least some dietary shift toward smaller, nonmigratory game—especially rodents and rabbits. In the long run the expansion of grasslands and open woodlands in the upper Rio Grande region may actually have increased the total availability of both animal and plant food resources, a condition that eventually would favor human population growth.

Modern anthropologists assume that the gathering of wild plant foods had always been a part of the lives of ancestral American Indians, even during the heyday of the megafauna and nomadic hunting. This assumption has been made if for no other reason than the fact that plant carbohydrates are known to have provided a valuable, if not essential, nutritional supplement in the diet of virtually every culture that has subsisted primarily on meat. Yet wild plants surely began to take on greater importance as a source for human food soon after the climate warmed and the vegetation and the fauna of north-central New Mexico underwent substantial change. In the words of noted paleoecologist Paul S. Martin, the setting was ripe for early hunters to "begin their 7,000-year experiment with native plants, leading to increasingly skillful techniques of harvesting and gathering, to the domestication of certain weedy camp-followers, and, within the last 1,000 years, to the widespread adoption of flood plain agriculture" (Martin 1963).

The early foragers probably continued to be highly mobile wanderers—procuring small quantities of plant and animal foods at a variety of locations where they weren't likely to have spent much time. In winter they must have tried to follow bison herds, possibly moving to the edge of the eastern plains. Small game and edible wild plants become scarce during the cold season, especially in the uplands, so these foods could not have sustained them. As yet there is no evidence that they stockpiled food.

Archaeological evidence for human habitation in the New Mexico Pueblo Province is scanty for the first thousand years or so following the eclipse of the great herding mammals. The kind of human-associated remains we might expect to have been left behind by small mobile bands are difficult to locate, subject to high erosion and poor preservation, and hard to date. The few sites that have been discovered tend to be primitive camps or game-kill sites yielding nothing in the way of decipherable plant remains or plant-processing tools. Stone crushers and grinding slabs likely used for pulverizing acorns or piñon nuts appear at later archaeological sites in the rolling country south and west of the Jemez Mountains. By the time the climate of the Southwest had somewhat stabilized once again, no later than 2000 B.C., the seeds of grasses and other wild plants were being processed as well.

Our comprehension of the fluctuations in climate and patterns of prevalent vegetation is clearer than our understanding of how the people lived during these early times. Recent studies of plant pollen grains and their association with datable soil strata have provided the basis for a fairly precise interpretation. For a long time following the wane of the great northern ice sheets, there were up-and-down swings in annual average temperatures and precipitation, as well as changes in the seasonality of precipitation.

However, a general trend toward a warmer climate was in progress during the first few thousand years following the glacial period. In northern New Mexico and southern Colorado a fairly wet climate seems to have prevailed at first. Frosts became less intense, and the main season for precipitation shifted from winter to summer. On the flanks of the mountain ranges spruce and fir forests retreated from lower to higher elevations and from south to north. Along the middle and northern Rio Grande Valley vegetation communities composed of piñon-juniper woodland and shortgrass mixed with shrubs gradually replaced ponderosa pine forests and tallgrass prairies. The ecosystem began to reflect the approach of conditions that continue to this day, conditions characterized by complex zonation of vegetation and pronounced seasonality. Indeed, except for the recent deteriorated state of native grasslands and the demise of a few animal species, the wild plant and animal life presently found in the back country of Bandelier, the Pajarito Plateau, and the outskirts of the Rio Grande Valley resembles the natural communities that prevailed for most of the past four to six thousand years.

From Collectors to Cultivators

IT WOULD BE A LONG TIME before the Pajarito Plateau would draw permanent residents. Human occupation during most of the so-called *Archaic period*, roughly from 5000 B.C. to A.D. 200, was mainly in the valleys and lower basins, by people who seem to have been on the move and may have camped briefly in rock shelters. Some of these rock shelters show increasing use over time. There's evidence for gradual population growth throughout the Archaic period, probably resulting from the increase of grasslands and open woodlands and a related rise in plant and animal diversity. The development of new techniques for processing plant foods also would have enhanced the success of each clan and tribe. Many Archaic archaeological sites dating to about 3000 B.C. are in places with relatively high botanical diversity in the immediate vicinity. We can assume that habitation sites were chosen at least partly because of convenient subsistence resources as well as nearby concentrations of potential food plants such as Indian ricegrass.

Edible plant species tend to be widely scattered, and their annual yield of fruits or nuts may be highly variable. Plants such as piñon pine produce nuts in quantity only every few years at a given location. Thus, individual sites aren't likely to have supported a population of any size or duration.

<small>PIÑON PINE
pg. 95</small>

The discovery by foragers of how simple hand-held cobblestones could be used over shallow-basined grinding slabs to smash wild plant seeds into a gruel or flour would have been a major technological breakthrough. Crushing plant material breaks open the cells and releases nutrients, achieving partial digestion much the same as cooking or chewing does. Archaeological sites in our area show that grinding implements began to be used there about the middle of Archaic times. These grinding stones were very heavy and were no doubt left from season to season as camps were seasonally occupied and abandoned.

Foraging implies a nomadic existence, offering little opportunity to stockpile food at any one location. As time passed, the people began the practice of keeping plant food by the process of drying or parching. Storing high protein, high carbohydrate food made it possible to live part of the year in permanent camps to which they would return on a seasonal basis. From these camps they could send out small hunting and gathering parties to places where they could obtain specific foods—piñon nuts and acorns in the fall, wild celery (*Cymopterus* spp.) or dock (*Rumex* spp.) in the spring. Gradually, their normal food-gathering routine shifted from foraging to collecting and storing the surplus.

<small>WILD CELERY
pg. 192
DOCK
pg. 169</small>

SLAB *METATE* AND ONE-HAND *MANO*, LATE ARCHAIC PERIOD (CA. 2000 B.C.–A.D. 300).

One grain plant, Indian ricegrass (*Oryzopsis hymenoides*), stands out from all others in its importance to the people. This native grass has long grown prolifically in central New Mexico, especially in the valley and basin soils of stabilized eolian sands associated with many of the archaeological sites of Archaic times. Whenever excavation research efforts have included ethnobotanical studies, parched ricegrass seed parts have turned up in high numbers at most of the sites of this era. The seeds are relatively large, are high in calories and protein, and ripen by mid-June—a time of year when protein would otherwise have been lacking in prehistoric diets. Ethnobotanical research indicates that other, larger-headed grasses such as dropseeds (*Sporobolus* spp.) and possibly New Mexico feathergrass (*Stipa neomexicana*), as well as the seeds from wild amaranths and goosefoot, were also used by early grain-processing societies.

There are indications of a general movement upward into montane habitats by late Archaic times, at least on a

INDIAN RICEGRASS
pg. 155

DROPSEED
pg. 160

NEW MEXICO FEATHERGRASS
pg. 160

AMARANTH
pg. 173

GOOSEFOOT
pg. 171

THE ANCESTRAL PUEBLOANS 19

NEW MEXICO FEATHERGRASS BELOW THE CLIFFS AT PETROGLYPH NATIONAL MONUMENT.

TINY CORN COBS EXCAVATED FROM JEMEZ CAVE ATTEST TO EARLY EXPERIMENTS WITH CULTIVATED FOOD PLANTS INTRODUCED INTO THE REGION FROM THE SOUTH.

seasonal basis. Here and there throughout central New Mexico and eastern Arizona, shallow caves show evidence of occupation during this period. But it was not occupation, per se, that has excited the archaeologists who have studied these sites. Among the yucca fiber sandals, willow baskets, and rabbit-fur blankets were tiny cobs and kernels of corn—the first direct proof of agriculture having made its way into the Southwest.

YUCCA
pg. 124
WILLOW
pg. 108

At first the dates assigned to the earliest evidence of corn grown north of Mexico were pegged at anywhere from 1500 to 3500 B.C. Later refinements in radiocarbon and other methods of dating, along with some reinterpretation of the original data, shifted the calendar for the initial appearance of cultivated plants considerably forward. Today, many archaeologists argue that corn and other cultigens did not enter the American Southwest until roughly 1000 to 1500 B.C., although evidence from as yet undiscovered sites may push these dates back somewhat.

One of the most important southwestern sites confirming early agriculture is Jemez Cave, located just above the upper piñon-juniper ecozone in the Jemez Mountains. Besides ancient corncobs that may date to 800 B.C., fragments of a cultivated pumpkinlike squash more than 2,000 years old have been excavated from this cave. Early fieldwork also revealed a plethora of wild plant remains, including thongs made of yucca, fragments of twined basketry from wild sumac branches, piñon nuts, and acorns. After studying the site in the mid-1960s, the renowned ethnobotanist Richard Ford concluded that "small groups of farmers occupied the shelter for a few weeks [each year] to plant and later returned to harvest whatever grew. Selective breeding [of corn] was not evident, competition with field weeds was permitted, and some years the corn was picked before maturity. Maize simply augmented their fall diet" (Ford 1975).

THREELEAF SUMAC
pg. 138

How did corn and other cultigens make their way into the Southwest? Corn is believed to have been domesticated from *teosinte*, a wild grass, in the semiarid highlands of central Mexico more than 7,000 years ago. Squash, beans, and other New World cultigens that would become important agricultural staples were also first cultivated and later domesticated in this region, but this occurred a bit later. The growing of corn and other crops spread through Mesoamerica and changed the way of life for its people. Yet agriculture did not spread north from Mexico and enter the American Southwest for many thousands of years, despite growing conditions that certainly were amenable to these annual plants. Why?

Currently there are two schools of thought. One is that the people living in this land would not have turned to agriculture until the stress of growing populations and depleted natural resources forced them to do so. The natural supply of wild plant and animal foods was no longer sufficient to support the burgeoning numbers of people. The other idea, put forth by W. H. Wills in his comprehensive book *Early Prehistoric Agriculture in the American Southwest* (1988), also assumes that the adoption of agriculture would have been delayed until population densities reached the point where it became an advantageous strategy. He proposes that the adoption of domesticated plants by foragers and collectors was a result of an *intentional* decision to reduce future environmental uncertainty, a form of insurance against subsistence failure. Wills and others believe that these people must have developed a means of storage for seed crops and may have made migrations twice a year to localities where cultivation was possible, once for planting and early tending and later for harvesting. Both theories may be valid and in combination may explain why it took so long for the people of the Rio Grande and other parts of the Southwest to adopt plant cultivation.

The introduction of agriculture seems initially to have had little immediate impact on the prevailing way of life. In the words of one prominent researcher, it was a "monumental nonevent" (Minnis 1985). The first use of domesticated plants is thought to have been a nondisruptive extension of hunting and gathering. Early fields probably were not given much care. Long periods of inattention must have been the rule at times when the people were collecting ripening wild plants elsewhere. Furthermore, the first corn to be grown in the Southwest had tiny cobs with only a few kernels per ear, not much of a resource to rely upon. Who could have envisioned what would come?

It may have taken more than a thousand years for gardening and farming to become a mainstay of the Rio Grande people. At first, corn and squash would have constituted only a small increment of their diet. But by the end of the Archaic period corn was being grown regularly on valley floodplains and had achieved a prominent role in the peoples' diet along with wild plant foods such as prickly pear, cholla (*Opuntia imbricata*), piñon nuts, groundcherry (*Physalis* spp.) fruits, various weedy annuals, and grass seeds. By this time, hunting had become less important to people living in the Rio Grande lowlands.

PRICKLY PEAR
pg. 189

CANE CHOLLA
pg. 140

GROUND-CHERRY
pg. 206

The end of the Archaic era was also a time when the montane regions on either side of the Rio Grande and its tributaries were becoming peopled by relatively large groups who occupied sites along cliff bases during fall and early winter. Along with a general increase in population, the trend from nomadism to seasonal occupation of sites, which later led to year-round dwelling, was accelerating. The stage was set for another way of life and a name for the people who adopted it—the Anasazi.

The Anasazi: Puebloan Ancestors

THE CHOICE OF THE NAME *Anasazi* for the people who lived in the northeastern part of the Southwest during the period between Archaic and historic times was unfortunate. Anasazi is a Navajo word with several translations, one of which is "ancestors of the enemy," a connotation that is an affront to the modern Pueblo Indians who are the Anasazi's

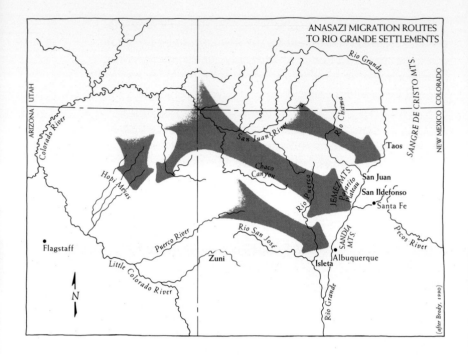

direct descendants. When it was first adopted by archaeologists in the 1930s, they understood it to mean "old people." Alas, the name has acquired nearly universal use in both technical and popular literature on the Southwest, and we retain it here as meaning "Puebloan ancestors."

The Anasazi era is considered by archaeologists to have started three to four centuries after the birth of Christ and to have ended when the Spaniards entered New Mexico. These ancestral Puebloans occupied the upper and middle Rio Grande drainage and most of the Colorado Plateau, centering in the country known today as "the Four Corners." In our area that era was characterized by greater use of the mountains and upland plateaus, continued population expansion, a gradual tendency toward a more settled way of life associated with a greater dependence on agriculture, and increasingly larger permanent villages and structures.

The Pajarito Plateau (*pajarito* is Spanish for "little bird") took longer to attract year-round inhabitants than many of the other upland areas in the Rio Grande region. In fact, it seems that within this era there was very little activity anywhere in the vicinity of the Pajarito Plateau between the Rio Grande and Rio Puerco river systems until after A.D. 1100. However, previous to that time the plateau and the Jemez Mountains above it were no doubt used on a seasonal basis for hunting deer and other game and gathering piñon nuts and oak acorns. Almost surely the uplands were visited regularly for the collection of numerous species of plants for medicinal use by tribes living in the surrounding lower country. But the pace of development varied widely from one region to the next, and it appears to have taken longer on the plateau, as it also did on the western flank of the Sangre de Cristo Mountains in the vicinity of modern-day Taos.

Immigrants to the plateau are believed to have come from sparse, semipermanent agricultural settlements scattered on the east side of the Rio Grande Valley. Then, or possibly a bit later, others came from the densely settled San Juan Basin to the northwest, where the highly developed culture of the people living in Chaco Canyon was breaking down. The Chacoans were sophisticated farmers, but their agriculture depended on late summer rains, which became less dependable in the late eleventh and early twelfth centuries A.D. When the people of Chaco left their fields and complex systems of dams and canals—to say nothing of their spectacular multistoried dwellings—and moved onto higher ground on the margin of the San Juan Basin, they had to adjust to a drastically new way of life.

Forced by drought to move up to the plateau and other highlands, farmers found an environment that had greater year-round moisture. But the growing season at higher elevations was shorter and the nights cooler, even in summer. These conditions were not conducive to relying upon agriculture to the degree the farmers had been able to at Chaco. Nor was the canyon-and-mesa terrain suitable for massive water-diversion systems. The people were faced with a choice: either to return to a seminomadic hunting and foraging existence or to establish settlements at sites conducive to small gardens for growing some corn and other crops and to rely heavily on wild plants and game within a day's reach. Since by now most of the people of the Southwest had long made a commitment to agriculture and a sedentary life, it was natural for the new residents of the Pajarito Plateau to opt for the second alternative. They relocated to the most promising sites for farming rather than revert to the old way of being always on the move. But because of fairly frequent swings in the prevailing climate, it would have been difficult for them to establish a stable food production routine, and this would be a time of relative poverty and off-and-on subsistence stress.

In the A.D.1100s the people who had moved to the Pajarito and into the mountainous areas to the north and west returned to living in deep pithouse structures. Pithouses were relatively easy to build and were thermally more efficient in the cooler climate than houses built above ground. Despite recent familiarity of these people with complex pueblo architecture, it would have made sense for them to expend minimum energy on construction of dwellings when they had to devote so much time to developing garden plots and collecting wild foods. In his delightful and informative book *The Magic of Bandelier,* archaeologist David Stuart put it neatly into perspective:

> This phenomenon is not really so surprising; nearly eight hundred years after dugout-pithouses were built high on the Pajarito Plateau, Anglo-American homesteaders in eastern New Mexico, Texas and Oklahoma were also building sod "dugouts" to live in until the first few crops had been planted and harvested. Only later did such "sod-busters" build the neat frame homes, courthouses and churches considered typical of Midwestern Anglo-American society (Stuart 1989).

On the central Pajarito it was not long before the people again began to build small, pueblo-type masonry dwellings, as they had during Chacoan times. This probably happened soon after their new fields became established and surplus corn could be stored for winter and early spring, seasons when most wild plants on the uplands are unavailable for food collection. Later, during a seventy-five-year period of generally stable weather and reliable summer rains, human population on the plateau began to increase, and habitation clusters were expanded to include more rooms for living and food storage. Eventually, crops from the adjacent fields were insufficient to feed the growing populations, and new fields a few miles from the widely spaced pueblos were cleared for cultivation. The people began to maintain dual residences—permanent village homes below and single-family field houses near the supplemental garden plots at higher elevations (Preucel 1988). A typical family might move to its outlying field in late spring; spend the summer engaged in tending crops, gathering wild plants, and trapping small game and birds; then return to the communal village with its harvests in the fall.

Since the fields were usually far from a stream, and dry farming for corn, beans, squash, and gourds was dependent upon summer rains, various structures were built to trap and hold moisture from the downpours that occurred sporadically from mid-June through September in most years. Stone check dams were constructed across arroyos to collect storm runoff, allowing fine-grained soil good for planting to build up behind the dams. Stone-lined reservoirs were built to hold water between one storm and the next. Other water-conservation devices associated with upland dry farming included rock terracing on steeper slopes and, on mesa tops and other flat areas, rock-outlined catchment systems (now known as *grid gardens*) that were sometimes paved on the bottom with a mulch of cobbles or gravel.

Mesa-top grid gardens became widespread in the uplands on either side of the Rio Grande and along the Rio Ojo Caliente during this period. Some of them occupied many acres. A typical grid garden might consist of a checkerboard of rectangles fifteen by twenty-five feet or smaller, each rock wall a foot or so high, with a dozen to hundreds of rectangles comprising a single garden. Rainwater would trickle off the walls into each grid, effectively increasing the amount of precipitation available for watering crops. Moisture levels were further enhanced by the use of gravel mulch, which acted as a barrier to evaporation. Sunshine reflecting from the rocks and solar radiation collected in the mulch caused the growing season to lengthen considerably. Scientists have speculated that a 120-day growing season may have been stretched to as much as 180 days. This would have allowed farmers to grow such late crops as cotton, which need a longer time to mature.

The production of corn required farmers to plant their crop in late spring and to bury the kernals deep—eight inches or more below the surface—where residual soil moisture from winter snows would have allowed germination and early growth. Still, periodic summer rains would have been necessary to bring the crop to maturity, about 120 days after germination, and in some years rainfall would have been insufficient.

Peter Pino, tribal administrator for Zia Pueblo for many years, has a keen interest in preserving traditional Puebloan ways. He recently undertook an experiment on one of the many long-abandoned grid garden plots that checker several of the mesas at Zia. The first year he planted corn kernels a few inches deep in spring, but no plants germinated. He decided the seeding wasn't deep enough, so the next year he seeded at an eight-inch depth. The corn sprouted and it grew to a height of about two feet; however, midsummer rains never came that year, and the plants withered and died. Peter hopes to try again in the future, but his first experiments do show how unreliable dry farming is today and would have been during Anasazi times, when the climate was very similar.

Intensively cultivated crops were not the only source of nourishment provided by ancient garden plots. When soil is disturbed, a habitat is created that favors various weedy annual plants over the hardy perennial flora that normally would thrive. Many of these native invaders are edible or have other human uses. Plants such as amaranth (*Amaranthus* spp.), goosefoot (*Chenopodium* spp.), purslane (*Portulaca* spp.), and mustards must have been relished by the Anasazi. It seems probable that the early tillers of these fields not only tolerated the useful "weeds" but may have encouraged them to grow among their intentional crops, as happens today in the old-style gardens preferred by some modern Puebloans. Rocky Mountain beeplant (*Cleome serrulata*), both a dietary staple and a useful plant for manufacturing black pottery paint, was a likely candidate for preferential treatment in prehistoric gardens. The frequency of beeplant pollen found in dried human feces (called *coprolites*) unearthed at several Anasazi sites has led some archaeologists to conclude that this "wild spinach" may have been extensively cultivated, perhaps the only nonintroduced plant in our area that was so manipulated for food production.

PURSLANE
pg. 178

ROCKY MOUNTAIN BEEPLANT
pg. 182

Additional benefits that the ancients surely would have derived from their active and fallow fields were the jackrabbits, cottontails, and occasional prairie dogs attracted to the fields, further supplementing the farmers' diets. And, no doubt, semidomesticated wild turkeys, which seem to have been associated with most every village of this era, found the fields and gardens to their liking, gobbling up insects and weedy annuals, turning summer plant food into rich animal protein for consumption by the people during the lean winter months.

During the late thirteenth and early fourteenth centuries A.D., another migration took place. People began to move from the rolling upland hills and higher mesas of the upper Pajarito Plateau down into the valleys, such as Frijoles Canyon in Bandelier National Monument, that contained perennial streams. This was a period of sporadic drought, which may have driven the people off their dry-farmed upland gardens and fields. During this Rio Grande Classic period, many of the great Anasazi pueblos were constructed,

THE ANCESTRAL PUEBLOANS **29**

ANCIENT GRID GARDENS ON MESA TOP AT ZIA PUEBLO.

usually not far from running water. The end of the fourteenth century and the beginning of the next mark the culmination of massive Anasazi pueblo and cliff-dwelling construction on the Pajarito—places such as Tyuonyi and Long House in Frijoles Canyon, Tsankawi and Otowi a few miles to the north, and Puye farther north. Other well-known examples are Tijeras and Paa-ko pueblos across the Sandia Mountains from Albuquerque, Kuaua, on the Rio Grande north of that city, and Arroyo Hondo Pueblo just south of Santa Fe.

Probably our best understanding of the cultures that thrived in these huge settlements, including their uses of wild and domesticated plants, comes from two decades of excavation and study of Arroyo Hondo Pueblo by the School of American Research (SAR) and the subsequent publication of eight major interpretive volumes on the prehis-

TYUONYI—ONE OF THE GREAT ANASAZI PUEBLOS, BANDELIER NATIONAL MONUMENT.

toric pueblo. Other large pueblos of the Rio Grande Classic period have been excavated, but none of the earlier research projects had so many sophisticated tools of modern archaeology at their disposal, nor did they bring in such wealth of expertise as was sponsored by the SAR.

The time span for the building and occupation of Arroyo Hondo Pueblo was relatively brief—a period of roughly forty-five years at the beginning of the fourteenth century A.D., and reoccupation and further building during another fifty-year period spanning the end of that century. At the height of development, around A.D. 1330, up to a thousand people occupied a two-story adobe pueblo consisting of twelve to fifteen hundred rooms. As with virtually all the Anasazi living in the Rio Grande drainage during this time, these people subsisted mainly on corn, also growing squash and beans. Located more than a hundred feet above

EXCAVATIONS OF ARROYO HONDO PUEBLO BY THE SCHOOL OF AMERICAN RESEARCH IN THE EARLY 1970S PROVIDED INSIGHT INTO THE ECOLOGY OF THE PEOPLE LIVING IN THE HUGE RIO GRANDE PUEBLOS DURING THE FOURTEENTH CENTURY A.D.

a fairly narrow, flat-bottomed stream course, this site provided limited acreage for irrigated farming. The settlers also practiced dry farming in small fields they had cleared in the piñon-juniper woodlands on the surrounding piedmont.

Still, except in the best of years, the cultivated fields within an hour or so from the main pueblo could not have provided nearly enough crops to feed a human population approaching a thousand. Archaeological evidence points to the additional harvesting of dozens of wild plant species that still grow within three miles of the pueblo ruins and whose remains turned up in many of the hearths and storage rooms excavated there.

It is estimated that in some years as much as fifteen percent of each Puebloan's caloric intake came from the seeds of weedy native annual plants growing around the fields, at the edges of the village, and on other disturbed areas (Wetterstrom 1986). When there was a good yield of piñon nuts, only about once every six years, food value from the harvest of this plant would probably have exceeded that from all other wild species combined. Fruits of the wild banana yucca (*Yucca baccata*) were another potential source of high-energy food.

In the eight-square-mile territory thought to have been the main growing and gathering area for Arroyo Hondo Pueblo, nearly half the plant species native to the area have edible parts. Most of these species were probably known to the people living there.

Nevertheless, there is plenty of evidence for regular cycles of stress from food deprivation at Arroyo Hondo. Tree-ring analyses reveal frequent droughts lasting two or three years. Signs of high infant mortality and remnants of plants considered to be starvation foods turn up every so often in excavated living and storage rooms.

Prickly pear and cholla are in the category of starvation foods. These cacti tend to produce juicy fruits in quantity every summer, even during times of exceptional aridity, and the fruits of some cacti remain on the plants and are edible well into the winter. In fact, remains of prickly pear fruits, known as *tunas*, turn up so regularly in coprolites at sites attributed to this era that one researcher has suggested that

PRICKLY PEAR CACTUS TUNAS.

the traditional corn–beans–squash food triad commonly linked with Anasazi diets would more appropriately be labeled corn–squash–prickly pear (Gasser 1982).

During times when corn and squash yields were low, the Anasazi may well have turned increasingly to other native seed and berry plants, especially grasses such as dropseeds and ricegrass and shrubs such as currants, gooseberries, and chokecherries—plants whose fruits are harvested in the fall by Puebloans to this day.

When drought became too severe and prolonged, as seems to have been the case near the end of the first occupation of Arroyo Hondo Pueblo, the people apparently migrated to small, perhaps temporary scattered sites nearer the Rio Grande. The second and final exodus from this pueblo was again related to a period of drought, which was followed by a catastrophic fire that destroyed most of the village.

The two departures from Arroyo Hondo Pueblo fit the pattern of other sites in the Southwest, where villages were dependent on a small water source and nearby land for dry farming. When populations living in these places exceeded sustainable use of the land, wild plant and animal resources within a day of the villages would quickly become depleted. Thus, even minor adverse changes in climate could induce enormous food-related stress. Furthermore, by the end of the fourteenth century A.D., the length of each period of drought had increased throughout the middle and northern Rio Grande region. Several of these periods lasted up to ten or more years (Orcutt 1991). Under these conditions it is not surprising that pueblos were often occupied for only a few generations before their inhabitants moved on.

By 1540 with the arrival of the first Spanish expeditions from the south into New Mexico—an event known as the *Entrada*, or entrance—almost exactly a hundred years after the final retreat from Arroyo Hondo Pueblo, most of the living pueblos of the Rio Grande were located near the main river or on one of its major tributaries. That last century had been a time of movement up and down the rivers, in and out of large dwelling complexes. By now the people were employing a variety of farming and water-development techniques as well as continuing their reliance on collecting wild plants both to supplement diets and to supply ingredients for medicines, dyes and paints, and implements and fiber products.

These people had learned to manipulate their environment so that their populations would multiply. They designed their villages to take optimal advantage of solar radiation and water retention. They wove decorated garments of cotton, dog hair, and wild-plant fibers, and they produced intricately designed pottery for both functional and ceremonial use. Their community organization was highly structured, and at least five different languages were spoken among the widely spaced communities. These are the ancestors of the present-day Pueblo Indians.

Suggested Reading

Axelrod, Daniel I.
1967 *Quaternary Extinctions of Large Mammals.* University of California Publications in Geological Sciences, Vol. 74. Univ. of California Press, Berkeley.

Betancourt, Julio L., Thomas R. Van Devender and Paul S. Martin.
1990 *Packrat Middens—The Last 40,000 Years of Biotic Change.* Univ. of Arizona Press, Tucson.

Cordell, Linda S.
1984 *Prehistory of the Southwest.* Academic Press, Orlando.

Stuart, David E.
1989 *The Magic of Bandelier.* Ancient City Press, Santa Fe.

Wetterstrom, Wilma
1986 *Food, Diet, and Population at Prehistoric Arroyo Hondo Pueblo, New Mexico.* School of American Research Press, Santa Fe.

Wills, W. H.
1988 *Early Prehistoric Agriculture in the American Southwest.* School of American Research Press, Santa Fe.

3

CONTACT, COLONIZATION, AND CATASTROPHE

When Gen. Francisco Vásquez de Coronado, his Spanish army, and his retainers clanked their way into New Mexico in the year 1540, they brought with them a band of one thousand stallions and probably an equal or greater number of cattle. For the most part, the rest of their provisions were obtained from the Indians, ostensibly by trading for trinkets. In actuality, the Spanish commandeered maize and other supplies as needed at the pueblos they encountered. In return, aside from a few metal tools, the army brought nothing of immediate value to the Puebloans.

It is thought that the progeny of the horses loosed from other advance expeditions and the breeding stock that followed became the mounts of Apaches, Utes, Navajos, and Comanches who later harassed some of the sedentary Spanish Mission and Pueblo farming communities, in some cases to extinction. Inadvertently, the animals may have left at least one little gift in their saddle blankets or their droppings—alfilaria (*Erodium cicutarium*), a garden weed from Spain that apparently arrived in the American Southwest via Mexico in the sixteenth century and proceeded to colonize New Mexico and Arizona. Alfilaria, also known as heron's bill because of its long, tapered seed, is now considered to be a good early-spring forage plant for sheep on dry rangelands. While a boon to herdsmen, the plant would have been useless to those without range animals. Other familiar weedy plants may accidentally have been introduced by European animals or as seed contaminating wheat or barley.

At least three thousand years before this encounter, the Indians of the Southwest had already obtained from Indians in Mexico the seeds and planting protocol for maize, squash, and, a bit later, beans, bottle gourd, and cotton. This protocol dictated the way different varieties were to be grown: when to plant (perhaps according to a local sundial), how deep, how far apart, and how many seeds to a hole or hill, along with local innovations to suit each particular microenvironment. Furthermore, over time, seed selection for replanting gradually resulted in varieties that grew better at more northerly latitudes than had the original Mexican stock.

To these, a wide variety of foreign herbal plants eventually were introduced. The European lamb's quarters (*Chenopodium album*) spread quickly at the expense of native plants, and even Mexican weeds hitched a ride northward. Purslane (in Spanish, *verdolaga; Portulaca oleracea*) quickly replaced its indigenous look-alike (*P. retusa*). In the journal of his attempted colonization of New Mexico (1590-91), Gaspar Castaño de Sosa spoke of the Indians having *quelites*, a word meaning edible greens and usually referring to weedy annuals associated with gardens such as pigweed and goose-

THE RIO GRANDE VALLEY PROVIDED THE CORRIDOR THROUGH WHICH THE EARLY SPANISH EXPEDITIONS MOVED NORTH INTO PUEBLO INDIAN COUNTRY.

foot. After the Spanish entered New Mexico, these native weeds were replaced by foreign analogues. With the introduction of at least a hundred nonindigenous plants, the flora of New Mexico was changed forever.

In 1540, during Coronado's explorations and depredations in the western part of the state, his chronicler Castañada described "a sort of oak with sweet acorns, of which they [the Zuni] make cakes like sugar plums with dried coriander seeds." The astonishing fact here is that coriander is a European spice (in Spanish the seeds are called *coriandro* or *culantro* and the greens are referred to as *cilantro*). How did it ever arrive at Zuni? Or did Castañada mistake coriander for stickleaf (*Mentzelia* spp.), one of the many native condiments familiar to the Zuni to this day?

Among the other familiar domesticated plants the Spanish encountered at Pueblo villages, and the subject of curiosity and debate, were watermelons and cantaloupe-like melons. Cantaloupe-like melons from Asian stock and watermelons originally from Africa were probably introduced into Spain by the Moors and brought to Mexico by Cortés twenty years before the *entrada* into New Mexico. Remarkably, seed from those same melon types appear to have been passed "hand to hand" or traded via Mexico to the Puebloans and brought into cultivation a step ahead of the Spanish arrival to the middle Rio Grande Valley.

At the end of the sixteenth century, Don Juan de Oñate gained the "right" from Spain to invade and organize the land of the Pueblos into a new Spanish kingdom. His contingent consisted of about 130 soldier-colonists as well as women and children, along with their baggage, commissaries, equipment, and tools. *Carretas*, bulky wooden carts whose wheels were made with cross sections of huge cottonwood trunks, provided most of the transportation.

Groaning, squeaking carretas, and the remains of a 7,000-animal herd of horses, donkeys, cattle, oxen, sheep, goats, and even pigs, and the entire entourage of colonists arrived at the doorstep of a northern New Mexico Tewa pueblo in the summer of 1598 (Hammond and Rey 1953). Although there is room for debate about the exact location and sequence of events, the Spanish apparently settled where the Chama meets the Rio Grande near present-day San Juan Pueblo. This location was far from the southerly Tewa Pueblo villages where earlier Spanish expeditions had spoiled any chance of a warm welcome by their misdeeds and excesses.

As it was, Oñate and his soldier-colonists presumed more than a bit when they established headquarters at San Gabriel/San Juan, displacing Puebloans whose ancestors had occupied the land there for some three hundred years. Still and however unwelcomed, Oñate ushered in an era of tremendous cultural exchange between the Spanish and the Pueblo people.

We are able to form a picture of the world Oñate entered from accounts of earlier expeditions, reports, letters home, and formal interviews by Mexican officials of those who had been to the northern province of New Mexico. Oñate's group would have encountered fierce cattle (bison) on the plains, but the Spanish had hopes of taming and interbreeding them with domestic cattle in order to improve their demeanor. The rivers teemed with fish, and there was plenty of wild game—especially deer and turkeys. Mountainsides abounded with edible acorns and wild cherry. The sierras were well timbered with pine, spruce, and fir and the woodlands with "cedar" and piñons. Grapes and plums lined narrow watercourses, and meadows and plains of grasses were so abundant with edible seeds that peoples could be and were sustained by them.

In their supply train of wagons and carts the colonists carried seeds for kitchen gardens and cuttings and seeds for orchards. (There is no record of grafting in New Mexico at that time; the cuttings were probably of grapevines, which could remain viable during the three-month journey from Mexico.) Both kitchen gardens of assorted vegetables and orchards of cultivated fruit trees were said to have been unknown among the Indians. Recorded eyewitness accounts after the initial colonization of northern New Mexico, as well as plant remains retrieved from excavations at the Palace of the Governors in Santa Fe and elsewhere, disclose a substantial list of European cultivars grown in New Mexico during the seventeenth and eighteenth centuries. The list includes wheat, coriander, apricots, wine grapes, peaches, pears, watermelons, cantaloupes, prunes, domesticated cherries, beets, carrots, cabbage, lettuce, onions, tomatoes, and radishes. Chile (*Capsicum annuum*), native to Mexico; hubbard squash, introduced into Mexico from Peru; and peas and lentils were also unknown to the Puebloans until they were introduced by Spanish, *Mestizos* (people of mixed Indian and European blood), and Mexican-Indian colonials.

At the time of the *entrada*, Indians of the Pueblo Province subsisted on varieties of maize, squash, beans, and gathered wild plants. They were prudent and tried to store enough food to last through a bad year. Communal hunts for rabbits, hares, deer, and other game provided their meat. They planted with fire-tempered digging sticks, tilled with deer scapula (shoulder blade) bones and stone hoes, and planted

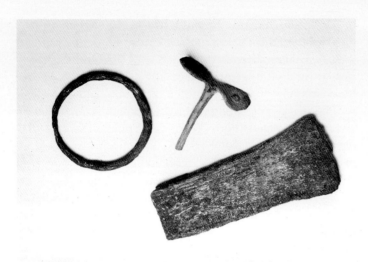

IRON TOOLS BROUGHT TO NEW MEXICO BY THE SPANISH COLONISTS:
FROM LEFT, **CHINCH RING, POT HANDLE, AXE, CA. A.D. 1540-1650.**

both in small, irrigated or flood-watered fields and stone-outlined mesa-top gardens. Their clothing was mainly of woven yucca and cotton fibers. Depending upon how far south their pueblo was, they either grew cotton or traded for it.

Timbers for their homes and kivas were hewed with stone axes and reused time and again. Firewood was taken from downed and dead tree limbs and shrubs. Although they were rich in family, culture, and land that supported all their needs, the Puebloans were basically a stone age people: they had no metal.

The colonists, on the other hand, brought wagons, draft animals, and iron tools: plowshares, axes, adzes, and hoes. They dug multivillage irrigation ditches and tapped upstream water so it would flow above their fields and could be manipulated where needed. They cleared and plowed large areas for fields and orchards. For beams, *jacal* (upright posts) structures, fences, and firewood, they cut forests and woodlands surrounding the villages and hauled the logs with draft animals. It wasn't long before herds of introduced domesticated animals grazed nearby on newly created pastures.

Soon after the Spanish colonists arrived in the Pueblo Province, disaster struck. Drought hit in 1601, and it hurt colonists and Indians alike. No rain fell to water the field crops such as wheat, barley, and maize. Water levels dropped in the rivers, drying up the *acequias*—irrigation ditches the colonizers had built with Puebloan labor upon their arrival. The recently planted orchards were too young to produce fruit. Hills near the village had been denuded, and the earth was trampled and hard-packed by cattle, sheep, and horses. When the rains finally arrived, they came as cloudbursts, eroding gullies in the fields and sheet-washing the parched land, which had no vegetation to hold back the torrents. The Puebloans' food reserve had long since been exhausted by the added demand placed upon it by the Spanish, and the Indians were just now learning to grow the new crops. Starvation appeared imminent.

This was probably the time when the Puebloans began to share their knowledge of wild plants with the settlers. They must have pointed the Spaniards to the mountains, where forests held the snowmelt longer and the land was largely immune to local drought. Nuts, berries, edible bark, and roots could be collected there. Indian knowledge of edible wild plants, acquired over the millennia, enabled the newcomers to last out the hard years.

This may also have been the time when the Spanish began to recognize wild plants that were familiar to them based on similar cultivated varieties back home. For example, osha (*Ligusticum porteri*) is related to the European lovage (*Angelica* spp.). Both herbs taste like celery and are used in the same way, as a green vegetable, a flavoring, and a medicine. Bee-balm, or *oregano de la sierra* (*Monarda menthaefolia*), with its square stems so characteristic of the many European herbs in the mint family, would also have seemed familiar to the Spanish.

BEE-BALM
pg. 202

Through nearly four centuries of good times and bad, the Spanish and Puebloans cultivated and consumed the

CONTACT, COLONIZATION, AND CATASTROPHE **43**

same foodstuffs. Among the Spanish contributions to the local cuisine were wheat, cultivated fruits, onions, and the chile that had been introduced to the Spaniards by the Mexican Indians. The Puebloans provided corn, beans, squash and wild fruits, nuts, vegetables, and herbs. Until recently the two cuisines have remained distinctive in terms of selection and preparation of ingredients. The Puebloan diet has always been rather simple, in contrast to the complexity of the culture itself.

Suggested Reading

Ford, Richard I.
1987 The New Pueblo Economy. In *When Cultures Meet*, papers from the October 20, 1984, conference held at San Juan Pueblo, New Mexico. Sunstone Press, Santa Fe.

Hammond, George P. and Agapito Rey
1953 *Oñate, Colonizer of New Mexico 1595-1628*, Vols. *1 and 2*. Univ. of New Mexico Press, Albuquerque.

Jones, Oakah L., Jr.
1979 *Los Paisanos—Spanish Settlers on the Northern Frontier of New Spain*. Univ. of Oklahoma Press, Norman.

Schroeder, Albert H. and Dan S. Matson
1965 *A Colony on the Move: Gaspar Castaño de Sosa's Journal, 1590-1591*. School of American Research, Santa Fe.

4
MODERN DESCENDANTS

In Harmony with Nature

NINETEEN PUEBLO INDIAN TRIBES continue to make their home in New Mexico in the region we are terming the Pueblo Province. Eight of them are located along the Rio Grande proper, eight along its flowing tributaries. Two other tribes, the Acoma and the Laguna, are located within the main watershed of that river but not on perennial streams. Zuni Pueblo is geographically within New Mexico but is west of the continental divide, on waters that flow into the Colorado River system. Two other Pueblo Indian tribes of the Southwest, the Hopi, located far to the west in Arizona, and the Ysleta del Sur near El Paso, Texas, are not covered in this book.

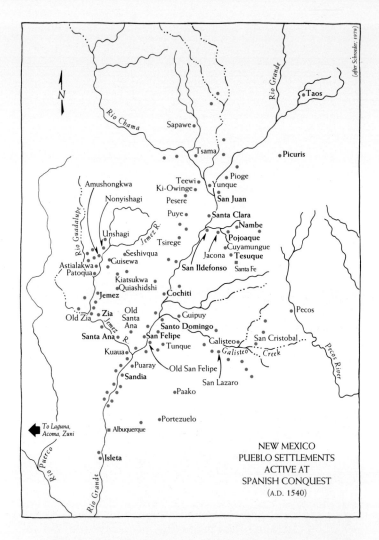

About 40,000 Puebloans now live in New Mexico, roughly the same number estimated to have occupied the land when the first Spaniards arrived in the sixteenth century. Five different Pueblo Indian languages are spoken. Social structure and customs vary considerably from pueblo to pueblo, as does religious organization. Linking all Pueblo cultures is an overarching view of life based on the long-ago emergence of a people who were, and are, one with the world and all its parts.

All things in the natural world—plants, animals, inanimate objects, even landscapes—are held to be sacred. Living in harmony with nature, respecting it, and sustaining it are central to the traditional philosophy and ecological orientation of American Indians. In effect, the land is an extension of Indian thought and being. In the words of Santa Clara Pueblo scholar Greg Cajete:

> The traditional relationship and participation of Indian people with the American landscape includes not only the land itself but the very way in which they have perceived themselves and reality. Indian people, through generations of living in America, have formed and been formed by the land. Indian kinship with this land—its climate, its soil, its water, its plants, and its animals—has literally determined the expressions of Indian theology (Cajete 1993).

ALL THINGS IN NATURE ARE HELD TO BE SACRED. CEREMONIAL CAVE, BANDELIER NATIONAL MONUMENT.

CREST OF THE SANDIA MOUNTAINS. MODERN DEVELOPMENTS HAVE NEGATED TRADITIONAL VALUES TO PUEBLOANS.

Dr. Cajete has been an American Indian educator for many years. He, along with many others, is concerned over the loss of lands that traditionally belonged to all the people and now are in private hands or are managed for "multiple use" (that is, commercial or heavy recreational use) by the federal government. Cajete observes:

> The pre-contact landscapes of America were as much an expression of Indian cultures as their arts and ceremonies. Today, the artifacts of Indian cultures are legally protected; yet the wellsprings from which such cultural expressions came—the land, the plants, the animals and the waters—are generally viewed, by mainstream society, as being outside the realm of cultural preservation (Cajete 1994).

Within the Rio Grande watershed, much land that once was the domain of native peoples has been lost through Spanish land grants or has come under the jurisdiction of the United States Forest Service, the National Park Service, or the Sandia and Los Alamos national laboratories. The traditional value of many once-sacred sites, such as the crest of the Sandia Mountains, has become lost forever to these people through either commercialization or development. Spiritually important sites continue to be threatened by government plans for new paved roads, power lines, high-impact recreational development, or other uses incompatible with quiet reverence.

Pueblo Indian philosophy has strong ties with the ethic of undeveloped wilderness, and a natural alliance often has resulted between the Puebloans and the environmentalists of modern urban society. Many of the latter are joining with their neighbors of the New Mexican pueblos to insist upon more enlightened national policies of land stewardship.

It is unfortunate that, in the name of free expression, a few individuals have revealed privileged information given by Puebloan friends in confidence. Likewise, it's unfortunate that other non-Indians have sought out ancestral shrines in which to act out their personal fantasies, perhaps believing they were usefully perpetuating old traditions or somehow learning what it means to be an Indian. Such people are an anathema to contemporary Puebloans.

In our contacts with Puebloans, we (the authors) have always made clear that we did not wish to learn about plant uses that should not be shared with outsiders. This book avoids detailing sacred or ceremonial uses of plants, except when this knowledge is meant for everyone's use, is conspicuous, or is otherwise already widely known by the public.

While knowledge of some uses of plants continues to be passed from one generation to the next, awareness of many other wild plant traditions is rapidly being lost, as younger Pueblo Indians are driven to keep pace with the modern world. A number of Puebloan educators and elders have expressed the wish that their children and grandchildren might learn about the plants that continue to be used today and that had such great importance to past generations. Toward this end we hope that, by compiling and presenting plant information specific to Puebloan use, we will have contributed to the preservation of a heritage.

Fields and Gardens

AS NOTED, THERE IS PLENTY of evidence that certain wild plants favor disturbed areas such as fields or gardens. These plants were incorporated into the diet of prehistoric peoples, and no doubt some were used for medicine as well. It's clear that this association has continued into modern times.

While he was a doctoral student at the University of Michigan in the late 1960s, Richard I. Ford undertook a comprehensive study of the ecological relationship between wild plants and the people of San Juan Pueblo. He learned that the farmers of San Juan categorized all plants growing in or around their fields either as useful plants or as weeds. Useful plants included planted crops, edible wild greens, and medicinal plants and were cultivated, encouraged, or at least tolerated. Weeds of no use to the Puebloans were hoed out. The farmers left an unplanted strip at the edge of their fields so useful species such as beeplant, purslane, and sunflower might thrive untended.

SUNFLOWER pg. 225

Ditch banks were another place where useful pioneering plants—such as blue trumpets (*Ipomopsis longiflora*), groundcherry (*Physalis* spp.), Indian tea (*Thelesperma megapotamicum*), and milkweed (*Asclepias* spp.)—were encouraged to take hold. Some people even transplanted certain medicinal and condiment plants—among them bee-balm (*Monarda menthaefolia*) and goosefoot (*Chenopodium* spp.)—to the ditch banks.

BLUE TRUMPETS pg. 198

INDIAN TEA pg. 223

MILKWEED pg. 196

In keeping with their respect for the elements in nature, the Puebloans practiced a strong conservation ethic when it came to utilizing wild plants. While they collected whole plants, a few in bloom were always left behind, permitting the population to perpetuate itself. Cactus stems collected for eating would be broken off at the joint so the rest of the plant might survive. If the men of San Juan needed some Apache plume (*Fallugia paradoxa*) for snares or brooms, only nonflowering branches would be taken. Amaranth or beeplant greens for the cooking pot would be broken off well above the soil line, allowing plant growth to continue.

APACHE PLUME pg. 134

When yucca roots were excavated for use as an ingredient in a natural shampoo, the collectors would then bury the seed pods, thus ensuring another generation of yuccas. Burying unused plant parts has another function: it permits the spirit of the plant to return to the spirit world.

From Ford's studies at San Juan, we know many heretofore unrecognized details about the relationship between the San Juan Puebloans and wild plants. More recent ethnobotanical research has confirmed that the threads of wild plant conservation run deeply among all Rio Grande Puebloans. An example of this ethic was informally demonstrated during a recent visit to Zia Pueblo. Tribal administrator Peter Pino spotted an old man, who was toting an even-older shotgun, resting alongside a back-country road. This fellow was employed by various members of the pueblo who own livestock to patrol the hills and mesas in order to protect against any wrongdoing and against wild dogs that might harass their animals. The old man talked about how he hunts rabbits and squirrels and collects various edible plants. He sometimes carries a pouch of Indian ricegrass seeds and broadcasts them about "because I just like nature, and this grass is a part of nature and there should be more of it around here."

In the last hundred years the people have been confronted with the reduction of traditional lands, the destruction of valuable farmland, and the introduction of a cash economy. Most agriculture practiced on Pueblo lands now is not much different from that found elsewhere in North America. Typically, large fields of monoculture crops are grown and harvested with all the latest machinery and techniques. But quite a few Puebloans still prefer to garden in the old way, cultivating a mixture of food crops such as corn, various squashes, beans, or chiles in a single plot, and tolerating, or even encouraging, some useful wild species among the crops.

In the course of researching this book, we photographed one particularly robust two-acre field at Zuni Pueblo. This garden consisted of dozens of patches and demirows of cul-

INDIAN TEA GROWING AMONG CORN ROWS IN AN OLD-STYLE GARDEN AT ZUNI PUEBLO.

LORRAIN LONCASION'S CONTEMPORARY WAFFLE GARDEN AT ZUNI PUEBLO.

WAFFLE GARDEN AT ZUNI PUEBLO, CA. 1935. VERY FEW WAFFLE GARDENS SURVIVE AT ZUNI TODAY.

tivated crops mixed with sunflowers, amaranths, purslane, wild potatoes, globe-mallows, and Indian tea. The owner of the field, David Chavez, told us he collects and uses some amaranth and Indian tea but usually doesn't have time to deal with the other plants. Some years, however, more of the "weeds" will be harvested.

WILD POTATO
pg. 210

GLOBE-MALLOW
pg. 187

Zuni Pueblo used to be famous for its "waffle gardens," so named because the texture of the surface resembles that of a giant waffle. In historic times these gardens usually consisted of a series of two- to three-foot rectangular compartments with clay earth walls a few inches high, built up to trap and conserve moisture. Very few waffle gardens survive at Zuni today, but in the main village one of them still features domestic onions and coriander, along with a number of wild groundcherry plants that volunteer, or seed in, each year. The owner of this garden, Lorrain Loncasion, said she encourages the groundcherries and mixes the fruits of this wild plant with her onions, coriander leaves, and chile to prepare a tasty salsa.

MODERN DESCENDANTS **53**

Some modern growers who practice old-style methods do so for other reasons than harvesting weeds. At San Juan Pueblo, Ramos Oyenque's field of corn, gourds, squashes, melons, and chiles appears weedy throughout. He says he likes the weeds, especially a wild red morning-glory that seems to cover everything. Because it serves as a green mulch for his cantaloupes and other low-vining crops, and gives early frost protection, the morning-glory is cheerfully allowed. Of course Oyenque and his wife also collect and cook up the amaranths, groundcherries, and purslane that abound among his irregular rows of "normal" crops.

Native Seeds/SEARCH is a Tucson-based nonprofit organization that works to preserve traditional native crops in the Southwest. Along the Rio Grande in recent years, this group has been working actively with the Puebloans to encourage old-style garden practices and to preserve ancient seed strains. In his fascinating book *Enduring Seeds*, the founder of Native Seeds/SEARCH, Gary Nabhan, neatly sums up an admirable philosophy:

> In essence, what many Native American farming traditions integrate with wild species within their cultivated fields and domestic economies is a dynamic balance of wildness with culture. This is what modern farmers lose when they cultivate their fields from edge to edge, leaving no hedges, no weeds, no wildlife habitat. The trend in industrial farming is, in fact, a repudiation of wildness. And yet, a certain wildness may be exactly what our ailing agricultural system needs.

A Multitude of Uses

PUEBLOANS HAVE LONG RECOGNIZED the small details that separate one plant species or group of species from another.

Nearly all important or common species have Indian names, which differ among the five Pueblo language groups in New Mexico. These names, which typically are compounded, refer to vegetative and flower attributes, including size, shape, color, feel, and smell, as well as plant habitat, associations with animals or humans, and the like. Medicinal plants are sometimes further named according to their ritual uses.

Just as the names for wild plants differ from pueblo to pueblo, so may the actual uses of them. Some plant uses seem to be nearly universal among the pueblos—the drying and brewing of hot beverages from the Indian tea plant, for instance. Individuals from virtually all pueblos still collect and brew this tea in much the same manner, as do the Navajo and people from other Indian tribes.

Uses of other plants, especially medicinal, may vary widely. Wild four-o'clock (*Mirabilis multiflora*), for instance, in the past has been used at Zuni to decrease the appetite, at San Ildefonso for indigestion, at Nambé for rheumatism, at Santa Clara for eye infections, at San Juan for swellings, and at Acoma for smoking tobacco. Even within a single pueblo, medicinal plant remedies may be employed in different ways by various households or clans.

FOUR-O'CLOCK
pg. 176

Before turning to species accounts of the most important useful plants, let's examine the major categories of uses, both prehistoric and recent, associated with wild plants of the Rio Grande.

Food and Beverage. Most of the evidence for prehistoric use of wild plants relates to food. Seeds contained in dried human feces, charred seeds in ancient hearths, pollen, and other plant remains unearthed from storage bins are all direct evidence of plants once eaten. Altogether, the archaeological evidence from these sources records more than twenty-five species of native plants used for food in the middle and upper Rio Grande region during prehistoric times (these are identified in the plant-use checklist beginning on page 254). The actual number, of course, is far higher, since many plant parts, particularly soft vegetation, do not preserve at all.

We've already discussed how the availability of wild plants and, later, the conditions for growing domesticated crops must have affected the settlement patterns of the ancestral Puebloans. We know that collecting wild plants for food has continued to this day, but as people became more and more dependent upon agriculture and as new crops were introduced into the Southwest, gathering gradually became more seasonal and less of a factor in the people's diet. Today, except for piñon nuts, most gathering for food by Pueblo Indians seems to be done in conjunction with other activities—tending domestic fields, hunting for game, or management of livestock—rather than of itself, yet more than 120 species of wild plants are known to have been collected and used as food or nonmedicinal beverages by Puebloans in New Mexico during the past century.

As corn became an ever-greater mainstay in the diet of ancestral Puebloans, the nutritional, rather than caloric, value of wild plant foods took on greater importance. Domestic beans, grown in the gardens and fields, are known to have furnished an important complement to corn, the two together providing an optimum balance of essential amino acids. Wild nuts and grains, cactus fruits, and amaranth and goosefoot seeds and leaves also are high in protein; their inclusion in early diets must have been critical to maintaining a healthy population.

Nutritionists are regularly coming up with new findings about beneficial nutrients contained in plant foods. For example, it turns out that purslane, once a vital ingredient of prehistoric Puebloan diets and still commonly collected for greens, is an extremely rich source of omega-3 fatty acids. Medical research during the past few years has demonstrated that these compounds tend to prevent oxidation of LDL, the "good" cholesterol that occurs in human blood, and so are helpful in combatting various coronary diseases. Purslane is one of a number of such wild plants that, although traditionally collected for the basic purpose of sustenance, actually provide additional hidden benefits.

In the age of overly processed food and ever-present chemical additives, it seems beyond dispute that supplementing with a few wild plant foods would benefit any diet. Certainly, many plants possess known or as-yet-unrecognized nutrients missing from processed foods. Many contemporary Puebloans, those who cling to some of the traditional ways, seem to know this intuitively, gathering and preparing wild celery and beeplant greens in the spring; amaranth greens, groundcherry pods, and wild onions in the summer; and piñon nuts and sunflower seeds in the fall. And nearly everyone still brews hot tea from the dried stems and flowers of the Indian tea plant, perhaps spiced with a sprinkling of wild mint leaves to take the chill off winter.

Medicine. Nowadays, wild plant gathering by Puebloans seems to be associated more with therapeutics than with food production. While nontraditional medicine provided at local hospitals, clinics, and drugstores probably accounts for most treatments, especially among younger Puebloans, the use of the old plant remedies continues, sometimes in a supplementary way, sometimes as the preferred way.

We can assume that many of today's medicinal plants were originally encountered by people seeking plants for new sources of food. It's not surprising that many wild plants likely to have been first ingested by experiment turned out to have therapeutic powers, for a high percentage of all plants occurring in nature contain chemical compounds endowed with curative qualities. In fact, more than forty percent of all "modern" medicines initially were derived from wild plant extracts.

For one example, at Zuni Pueblo fourteen different plant species are used in the treatment of stomachache (Camazine and Bye 1980). Nearly all of these contain chemicals with emetic qualities, such as saponin in the wild buckwheats or croton oil in doveweed. Saponin irritates mucous membranes and can cause vomiting when ingested. Croton oil

WILD BUCKWHEAT
pg. 167

DOVEWEED
pg. 185

contains a number of acids that irritate the gut, also causing nausea and vomiting.

Altogether, some 180 different species of wild plants growing in the Rio Grande watershed have medicinal uses associated with one or more of the nineteen Pueblo Indian tribes of New Mexico. Some are used in strictly ritual ways and aren't mentioned in this book. Others possess cures based upon a combination of ritual and chemical properties. Other remedies come from pharmacological agents derived from plant parts.

Tannic acid is one such agent. Modern medicine scientifically explains what the Puebloans knew from practical experience: tannic acid causes the precipitation of protein in skin and gastrointestinal tract cells, forming a protective barrier that can be helpful when tissue is injured or stressed. A number of wild plants contain high amounts of tannin, and several of them—wild dock, puccoon (*Lithospermum* spp.), wild geranium, and willow bark—have been used by Puebloans to treat sore throats, skin abrasions, and rashes. The people may not originally have been aware of the chemistry of these plants, but through trial and observation they found them to be effective as medicine for mild skin and throat irritations.

WILLOW
pg. 108

But the chemical constituents of most wild plant species have yet to be analyzed. The presence of effective chemical compounds is suggested but not scientifically confirmed for many plants linked with a medical tradition. Thus, modern pharmaceutical companies maintain a persistent interest in ethnobotany.

Then there are the healing recipes calling for a mix of herbs, a mix likely to have evolved from family traditions. We observed the practice of this kind of informal medicine during a recent visit with Jemez Puebloan Patrick Toya the morning after he and his family had returned from a long weekend of manning a food booth as volunteers at the annual Indian Market in Santa Fe. During our visit, Toya's adult son brought in a small bundle of wild plants he had just collected from the hill behind their home. The bundle consisted of branches of juniper, wild buckwheat (*Eriogonum* sp.), feather dalea (*Dalea formosa*), and a sprig of piñon pine. Still feeling the effects of the hectic weekend, the young man said he was going to brew a tea from these wild herbs to clean himself out. Sure enough, all four plants are known to have been used by Puebloans for medicinal purposes at one time or another, but perhaps only the Toyas and a few other families have come up with this particular combination.

JUNIPER
pg. 105

Some of the medicinal plants associated with Pueblo people grow only at higher elevations. Not many of these are detailed in this book, since most of them can't be seen along the park trails we cover. But certain members of the medicine societies that use these plants still make regular forays to the mountains to collect them. Thus, many Puebloans are concerned about having continued legal access and rights to use these lands for medicinal herb collection, even though they may technically be outside the official boundaries of their pueblo. These people also have a legitimate interest in such lands, which typically are within the national forests of New Mexico, not being developed for heavy recreational or commodity uses that might diminish the quality of habitat for their medicinal plants.

Baskets, Blankets, and Twine. Clay pots have been manufactured in the upper Rio Grande basin only during the past fifteen centuries or so; before then, animal skins and wild plant fibers were the only sources of materials for making baskets, water jars, and other containers. Fragments excavated from archaeological sites show that yucca leaves, split lengthwise into long strips, were the preferred fiber for woven vessels, with beargrass (*Nolina microcarpa*) running a distant second at sites where it was more readily available. Branches from willow or threeleaf sumac were employed in constructing coarser baskets or trays, especially the coiled type. A few artisans, such as Corina Waquie of Jemez Pueblo, still make and sell baskets from native material. Waquie prefers to use narrowleaf yucca and says that supply

BEARGRASS
pg. 127

THREELEAF
SUMAC
pg. 138

NARROWLEAF
YUCCA
pg. 124

MODERN DESCENDANTS **59**

ABOVE
STRIPS OF YUCCA LEAVES SPLIT LENGTHWISE WERE THE PREFERRED FIBER FOR MAKING WOVEN VESSELS DURING THE ANASAZI ERA.

RIGHT
YUCCA SANDAL, CA. A.D. 500-700.

SAN ILDEFONSO BASKET MADE FROM PLAITED WILLOW BRANCHES, EARLY 1900S.

is getting limited since it can't be collected on national forest lands anymore.

For coarsely woven sleeping mats, cattail leaves were sometimes combined with yucca strips. Blankets were woven from finer yucca fiber that may have been laced with turkey feathers or rabbit fur, and these would have served as shawls or robes by day and have provided warm, soft bedding at night.

CATTAIL
pg. 153

The introduction and cultivation of domestic cotton, along with the acquisition of knowledge about looms, occurred in our area less than 1,300 years ago. Before that time weavers of textiles or yarn had to rely on human hair or the hair or wool from domesticated dog, wild animals, bird skins or feathers, or leaf and stem fibers from wild plants. Fiber plants included milkweed and dogbane (*Apocynum* spp.) and, of course, yucca and beargrass. For a fascinating look into the preparation of prehistoric fibers, yarns, and dyes, Kate Peck Kent's book *Prehistoric Textiles of the Southwest* is highly recommended.

Probably the single most important ancestral use of fiber was in cordage. Yucca leaves were boiled, then beaten or chewed, and finally soaked to extract their fibrous strands, which were twisted together to form a kind of twine or rope. Besides being used as a binder for most of the household articles noted above, yucca cordage was employed in making fishnets, rope ladders, head straps, thongs for sandals and snowshoes, belts, and headdresses.

Dye, Paint, and Tanning. When the Spaniards first marched up the Rio Grande Valley, they must have been surprised to find the Pueblo Indians dressed in embroidered cotton garments decorated with figures in many colors (Minge 1979). Locally obtained native plants were most certainly the principal source for pigments. Although little research has been done to evaluate chemical evidence that would indicate which dye plants were used along the Rio Grande in prehistoric times, we can postulate on this subject based upon historic use here and throughout the Southwest. More than twenty species of native plants growing in our area have been used by American Indians for textile dyes (Tierney 1977). The most commonly used plant parts include mountain-mahogany (*Cercocarpus montanus*) roots and alder bark for red dye, especially for tinting buckskin; rabbitbrush (*Chrysothamnus nauseosus*) blossoms and whole Indian tea plants for yellow; and the twigs and leaves of threeleaf sumac mixed with piñon pine gum to produce a good, solid black.

MOUNTAIN-MAHOGANY
pg. 136

RABBITBRUSH
pg. 148

Plant dyes also have been popular for coloring certain foods. Fourwing saltbush ashes were used to tint wafer bread a greenish blue, ground amaranth seeds to impart a red stain to Zuni piki bread, and the red tunas of the prickly pear to color cornmeal mush.

FOURWING SALTBUSH
pg. 129

The earliest pottery manufactured in our area, around A.D. 400, was undecorated, but later a black, mineral-based pigment obtained from iron-bearing clay was employed. It was not until about A.D. 1225 that black organic paint was used on ceramics made in the Rio Grande Valley (Dittert and Plog 1980). A rich black was produced from a concentrate prepared by boiling the leafy stems of the Rocky Mountain beeplant, which contains high amounts of iron. Painted designs would typically be applied after the pot was slipped. Upon firing, the color would soften to a dark gray. West of the Rio Grande, tansy mustard (*Descurainia* spp.) was often used instead of beeplant, but in much the same way, to prepare black pottery paint. Other wild plants associated with pottery painting include wild dock and piñon pine. It probably is no coincidence that all these plants known to produce ceramic pigments also happen to have edible parts, which leads to speculation about how their use as colors was first discovered.

TANSY MUSTARD
pg. 180

Rock-hugging lichens are another group of plants used to produce paint. When ground and mixed with piñon resin, a deep yellow hue results. Certain lichens also had medicinal applications.

Beeplant is still used to produce black pigment by some modern Pueblo potters who cherish traditional methods. Acoma potters are especially likely to prefer painting the finest lines with a brush made from the chewed ends of yucca leaf strips.

Oily deer brains are thought to have been the main ingredient the early Puebloans used in tanning skins or hides. Later, wild plants were used, as described by Ruth Underhill in her authoritative little book, *Pueblo Crafts*:

> After the Spanish came, Pueblo people learned a vegetable tanning method, but they suited it to their own conditions and to the plants they could find. The European way was to boil powdered oak bark, and soak the skins in vats of this liquid for two or three weeks. Pueblo people who could not often get oak, used white fir (*Abies concolor*), canaigre (*Rumex hymenosepalus*), or Mormon tea (*Ephedra nevadensis*). They dried the bark of the trees or whole stems of the plants in the sun, then boiled and pounded them to powder. They mixed two parts of this liquid with one part of water and

WHITE FIR
pg. 102

MORMON TEA
pg. 122

MODERN DESCENDANTS **63**

BEEPLANT, HIGH IN IRON, WAS THE SOURCE OF THE DARK-COLORED PAINT ON THIS SAN ILDEFONSO POLYCHROME POT, CA. A.D. 1770-1790.

soaked a skin in a pot or other clean container which had no metal about it, airing the skin each day so that the acid would not eat into it. Finally it was rinsed in clear water and hung up to dry.

Construction. Whether it was a single-dwelling pithouse or a multistoried pueblo, construction always involved combining earth and wood. The latter was used for support beams and roof *vigas* (roof cross beams) as well as roof thatching. The critical thing for shaping good beams was finding trees with straight, relatively unbranched trunks. However, piñon pine and one-seed juniper—short, gnarly trees rarely suitable for making the heaviest beams—are the most common species growing near prehistoric village sites on the lower slopes of the Pajarito and other plateaus and hills bordering the Rio Grande. (Both piñon and juniper often were employed in lighter roof work.) Instead, wood from the clear, straight boles of ponderosa pine, Douglas-fir, and cottonwood would have served best as support structures, the first being the preferred tree for heavy construction at most Frijoles Canyon sites. Since these larger trees grow predominantly in forests located above, and sometimes miles away from, many of the principal villages, great effort must have been required to cut and haul the timbers.

PONDEROSA PINE
pg. 99

DOUGLAS-FIR
pg. 102

COTTONWOOD
pg. 111

This must have been especially true for the large population living in Chaco Canyon in the tenth century A.D. Although more trees would have been growing in the vicinity than are found there today, Chaco was hardly a densely timbered area at the time the multistoried Pueblo Bonito was being constructed. Yet of the five huge beams supporting the roof in one second-story room, four different species of

FOUR DIFFERENT SPECIES OF TREES WERE USED FOR THE CEILING BEAMS OF THIS THOUSAND-YEAR-OLD SECOND-STORY ROOM AT PUEBLO BONITO IN CHACO CANYON. A FIFTH WILD PLANT, COMMON REED, WAS LAID ACROSS THE BEAMS FOR CLOSURE. PHOTOGRAPH TAKEN PRIOR TO STABILIZATION OF THE RUINS.

MODERN DESCENDANTS

COMMON REED pg. 158

trees have been identified: ponderosa pine, Douglas-fir, cottonwood, and juniper. Common reed (*Phragmites communis*) was employed as the closing material in this particular ceiling. Although the cottonwood, juniper, and reed could have been obtained locally, the ponderosa and Douglas-fir tree trunks probably had to be hauled in from the Chuska Mountains, fifty miles west of Chaco.

OAK pg. 114

Smaller branches of oak, mountain-mahogany, and fourwing saltbush, along with piñon and juniper, also have been identified in roof structures at many Anasazi sites. It seems likely that parts of these shrubs and trees were used as *latillas* (wood lath), as well as for closing material.

Implements. Most of the woods mentioned above were also used in one way or another to craft various prehistoric implements. Arrows, of course, required straight pieces of tough wood. The branches from threeleaf sumac, oak, wild currant, mountain-mahogany, fourwing saltbush, and Apache plume meet these criteria, and all these shrubs are known to have been used in this way by ancestral Puebloans. When it was available along the river, common reed was also sought out for arrows because of its straight shaft and light weight. Shafts made from common reed would usually be tipped at either end with one of the above harder woods for the arrow point and the bowstring groove.

CURRANT pg. 132

NEW MEXICO LOCUST pg. 120

For bows, more pliable wood was needed; New Mexico locust (*Robinia neomexicana*), wild currant, and chokecherry were commonly used. Glue for attaching feathers to arrows and sinew to bows, and for many other adhesive jobs, came from pitch boiled down from piñon pine. Piñon gum is still occasionally used as a substitute wood glue by modern Puebloans.

There is no strict association between plant species and their use for many of the wooden objects made in the past. Depending on the pueblo, various woods seem to have been used for digging sticks, throwing sticks, cradleboards, tool handles, embroidery stretchers, and the like. Other plants appear to have had more universal specific associations—the use of Apache plume for making brooms, for example, or the highly revered Douglas-fir for prayer sticks.

Drums, Decorations, and Petroglyphs. The mesmerizing beat of the drum accompanying a small knot of male chanters is an integral part of nearly all Pueblo dance ceremonies. Drum making is still practiced by craftspersons of several pueblos, but nowhere is it more developed and important to society than at Cochiti Pueblo, where forty-foot-tall water reservoirs, brightly painted to portray giant drums, mark the center of this community. Bill Martin, one of the renowned Cochiti drum makers, uses two different native trees for the wooden parts of his drums, as do most other drum makers today. He prefers aspen (*Populus tremuloides*) wood for small and medium-size drums, since aspen tends to rot in the center, saving him time when he hollows the core of the wooden cylinder. Bill says that although pine logs were sometimes used in the past for large drums, their shells were too heavy. He believes that aspen was *the* traditional wood for drums at Cochiti because of its lightness and ease in burning out the center in days when effective chiseling tools were lacking. Cottonwood, the better-known traditional material, tends to crack more, but because of its size must be used for his largest drums. Most modern drum makers have to buy aspen imported from Colorado or farther north, but in times past they would have journeyed high into the Jemez, Sangre de Cristo, or Sandia mountains to obtain it, or have gone down to the river for cottonwood.

ASPEN
pg. 111

In the minds of many, it is the crafting of exquisite pottery that best symbolizes Puebloan art during historic times. The depiction of wild plants or plant parts as decorations on pottery or on other objects, both prehistoric and recent, is rare, and its near-absence is somewhat puzzling. For instance, when thousands of potsherds were examined during the excavation of Pecos Pueblo between 1915 and 1927, only four fragments could be said to have representations of plant parts. Likewise, an earlier survey of the pottery of the Pajarito Plateau revealed no floral decorations whatsoever

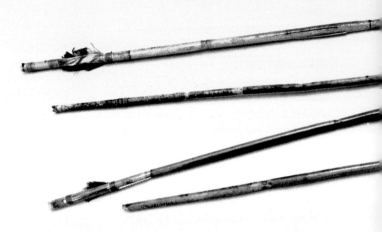

(Kidder 1915, 1932). When naturalistic plant designs have been used, they usually depict corn or other cultigens rather than wild plants.

A lack of wild plant images also holds true for the murals on some of the interior walls of excavated kivas. When plants such as yucca are portrayed, they tend to be shown as implements made from the plant, not the plant itself.

Nor do petroglyphs exhibit many examples of wild plant forms. Notable exceptions include two prominent incised images, of what appear to be yucca seedpods, that can be seen from the Macaw Trail in Boca Negra Canyon at Petroglyph National Monument. At Bandelier National Monument the only petroglyph known to incorporate an identifiable native plant in its motif is found on the cliffs at Tsankawi. This one seems to show a young white fir or Douglas-fir among some kachina figures (Rohn 1989). (Most contemporary Puebloans believe that the ancient petroglyphs are representations of sacred symbols rather than expressions of art; thus some feel that the widely used term "rock art" is inappropriate.)

We can only speculate as to why wild plants were rarely depicted on pottery or in murals or petroglyphs.

ARROWS CRAFTED FROM COMMON REED, CA. A.D. 700-1500.

MODERN DESCENDANTS **69**

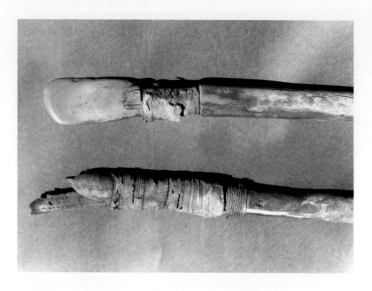

ABOVE
ONE-THOUSAND-YEAR-OLD DIGGING STICKS FROM CHACO CANYON. THE UPPER STICK IS TIPPED WITH CARVED BIGHORN SHEEP HORN FASTENED WITH YUCCA CORDAGE AND ANIMAL HIDE; WORKED STONE TIPS THE LOWER STICK. TIPS OF HARD MATERIAL ALLOWED LIGHTWEIGHT COTTONWOOD TO BE USED FOR THE HAFTS.

RIGHT
CRADLEBOARD FROM COCHITI PUEBLO, CA. EARLY 1900S. THE BOARD IS CRAFTED FROM PONDEROSA PINE AND THE HOOD HOOPS FROM THREE-LEAF SUMAC.

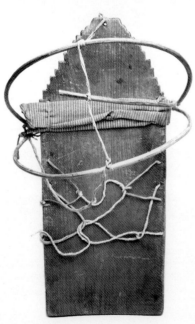

COTTONWOOD DRUM FROM COCHITI PUEBLO, CA. EARLY 1900S.

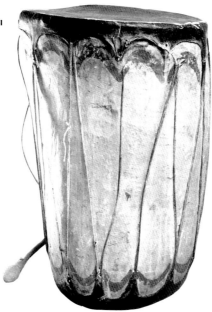

PETROGLYPH DEPICTING YUCCA SEEDPOD, PETROGLYPH NATIONAL MONUMENT.

MODERN DESCENDANTS 71

Perhaps plants did not in this manner capture the imagination of the people who created these designs. Or perhaps there was some taboo against their use.

Suggested Reading:

Cajete, Gregory
1994. *Look to the Mountain—An Ecology of Indigenous Education.* Kivakí Press, Durango, CO.

Ford, Richard I.
1992 *An Ecological Analysis Involving the Population of San Juan Pueblo, New Mexico.* Unpublished Ph.D. dissertation, Garland Press, New York

Kent, Kate Peck
1983 *Prehistoric Textiles of the Southwest.* School of American Research, Santa Fe, and Univ. of New Mexico Press, Albuquerque.

Nabhan, Gary Paul
1989 *Enduring Seed: Native American Agriculture and Wild Plant Conservation.* North Point Press, San Francisco.

Peckham, Stewart L.
1990 *From This Earth: The Ancient Art of Pueblo Pottery.* Museum of New Mexico Press, Santa Fe.

Rohn, Arthur H.
1989 *Rock Art of Bandelier National Monument.* Univ. of New Mexico Press, Albuquerque.

Sando, Joe
1991 *Pueblo Nations: Eight Centuries of Pueblo Indian History.* Clear Light Publishers, Santa Fe.

Underhill, Ruth
1979 *Pueblo Crafts.* The Filter Press, Palmer Lake, CO.

5
INDICATOR PLANTS AS LIVING ARTIFACTS

*A*S YOU OBSERVE THE NATURAL SETTING of ancient ruins such as those at Bandelier, Jemez, and Coronado, you may wonder why the patterns of vegetation growing on and around these places often appear different from those found on adjacent land. These differences have also piqued the curiosity of professional archaeologists and botanists, who developed the concept of *indicator plants*, that is, native or introduced plant species whose nature, distribution, and abundance correlate to specialized local soil conditions often associated with the past activity of indigenous people or early settlers.

The observation of indicator plants has been motivated by the wish to find answers to many questions. Can surface vegetation suggest the best place to dig a shallow well? Does plant growth ever reveal the presence of valuable underground mineral resources? Did the occurrence of certain native trees or shrubs give early homesteaders a clue to where good, deep farming soil might be found? Are certain unique plant assemblages a sign of the former presence of humans? The answer to these kinds of questions more often than not is a qualified *yes*, but the reasoning behind that answer is seldom direct and often rivals the sleuthing of a Sherlock Holmes.

One of the most well-known examples of an indicator plant is cottonwood (*Populus* spp.), whose growth in bottomlands invariably indicates the presence of a relatively shallow source of groundwater. Less obvious are the associations of antelope-sage (*Eriogonum jamesii*) and milkvetch (*Astragalus* spp.) with uranium ores, or the occasional association of mesquite with petroleum-bearing rocks. Historically, early settlers of the West noted that big sagebrush (*Artemisia tridentata*) indicated deep, fertile soils; usually, the taller the plants the deeper, better structured, and less saline the soils. With sufficient rainwater and a little luck, these areas were considered well suited for marginal dry farming.

ANTELOPE-SAGE
pg. 167

SAGEBRUSH
pg. 150

The introduced white clover, which became known as "white man's foot," is another indicator plant. To the American Indian its presence not only heralded the intrusion of non-Indians into their territory but also the heretofore unknown stinging honeybee. This led Longfellow to pen:

> "Wheresoe'er they tread, beneath them
> Springs a flower unknown among us,
> Springs the White Man's Foot in blossom"
>   ("Hiawatha," 1855)

Perhaps the first published scientific recognition of indicator plants was recorded by that keen observer of natural history, Charles Darwin. In 1832, while exploring Tierra del Fuego during his voyage aboard *The Beagle*, Darwin noted that shell heaps of former Fuegian inhabitants could be "distinguished at a long distance by the bright green colour of certain plants, which invariably grow on them. Among these may be enumerated the wild celery and scurvy grass, two very serviceable plants, the use of which has not been discovered by the natives."

About 120 years later and a continent away, Swedish scientist Olof Arrhenius attempted to determine prehistoric settlement patterns by the relative amount of phosphorus in the soil. From his previous work in Sweden, Arrhenius knew that elevated levels of phosphorus are often coincident with organic material remains at old habitation sites, and he speculated that he would also find high phosphate levels in the soil around ancient villages in New Mexico. Soils from a buried Indian ruin at Bandelier and from other prehistoric

WOLFBERRY AND FOURWING SALTBUSH GROWING BELOW CLIFFS AT TSANKAWI, BANDELIER NATIONAL MONUMENT.

sites near Santa Fe were sampled, but there was little correlation with phosphate levels, and, generally, his results were inconclusive (Arrhenius 1963). Elsewhere, high phosphorus levels have usually been associated with prehistoric sites containing bones and evidence of animal butchering. Unintentionally, Arrhenius corroborated the fact that the ancient Pueblo Indian diet wasn't meaty, something many archaeologists and historians have known for a long time.

The ruin mounds that Arrhenius studied still stand out from the surrounding hills. They are covered with wolfberry (*Lycium pallidum*), fourwing saltbush (*Atriplex canescens*), cane cholla (*Opuntia imbricata*), or, where the ruins have been recently excavated or pot-hunted, summer cypress (*Kochia scoparia*). A little botanical or ecological know-how has redeemed many a student of archaeology who was able to discern old ruins from a distance through the telltale vegetation growing on the surface around them.

In areas of low nitrogen, such as arctic tundra, the impacts of humans may be more direct. A vegetational study of a 3,600-year-old Eskimo site at Walakpa Bay in Alaska brought this out (Potter 1972). Sodhouses were used at this site as recently as a hundred years ago. Grass cover of the slopes from the eroded sod walls was determined to be up to ten times greater than cover on the surrounding tundra, and a highly significant increase in bluegrass (*Poa* spp.) and liverwort (*Marchantia* spp.) occurred on the inland (leeward) side of the houses—environmental evidence of nitrogenous soil enrichment. Eskimos frequently tanned skins against the sides of their houses, using nitrogen-rich urine in the tanning process. They were also most likely to have urinated on the leeward side of their dwellings, never giving a thought to the long-term effects on local vegetation.

The occurrence of indicator plants on old sites of human habitation reflects conditions favorable to the growth of these plants. Soil acidity, soil phosphorus, soil texture, water availability, and subterranean compaction are factors that have been locally and (for the plants) favorably altered by the past activities of humans.

WOLFBERRY
pg. 143

The wolfberry shrub is concentrated over buried prehistoric rooms at archaeological sites in some areas. Ecologist Loren Potter and his colleagues have explained this phenomenon initially through observing that wolfberry is favored by relatively moist soils, such as occurs between subterranean rocks. Where wolfberry was aligned in rows, buried rock walls of Anasazi sites were found below the surface. Where greasewood (*Sarcobatus vermiculatus*) was growing in angular patterns, and taller and greener than in surrounding areas, the cause was determined to be percolating rainwater, trapped by the artificial catchment basin of the compacted floor of an ancient dwelling (Potter and Young 1983).

When he conducted his investigations at Chaco Canyon, Potter would note any anomalies in species, alignment, or vigor of wolfberry or other indicator plants. From these occurrences he was successfully able to predict the presence of buried ruins, even though there was no other physical evidence of what was interred below.

Perhaps this phenomenon of buried catchment basins points to the former presence of Archaic peoples in the sandy terrain of the Rio Puerco drainage west of Albuquerque. It has been suggested that wolfberry shrubs sometimes found there indicate that archaic seasonal shelters with packed dirt floors were constructed long ago on the native Indian ricegrass and sand dropseed prairies.

Western tansy mustard (*Descurainia pinnata*) also tends to thrive above prehistoric buried rooms, owing to the increased clay content of the soil, a result of the weathering of former adobe walls (or clay mortar used on stone walls). Increased clay is also implicated as a factor in the *absence* of broom snakeweed above buried rooms. Although snakeweed is a common shrub on various soils, especially the overgrazed rangelands that often surround prehistoric sites, it seldom grows on the higher-clay soils associated with the sites themselves.

Cane cholla is still another plant that may indicate old habitations, especially where it occurs clearly out of its natural range. For example, cholla has been found growing on

BROOM
SNAKEWEED
pg. 145

archaeological sites in ponderosa pine habitat north of Jemez Pueblo, at a higher elevation than where it normally occurs. It has been suggested that the stands of cholla north of the pueblo are descendants of plants brought to that area for food or fencing between A.D. 1250 and 1700 (Housley 1974).

In the Pueblo Province the frequently observed concentration of fourwing saltbush on mounds covering prehistoric trash middens may be a counterpart of the phenomenon Darwin observed at Tierra del Fuego. In his survey of numerous sites on the Pajarito Plateau, archaeologist Charles Steen noted that at one location a stand of fourwing saltbush grew only on one mound. At the 6,900-foot elevation of this particular ruin, saltbush is not normally found. Steen speculates that the saltbush plants here could have survived on the rich soil of the midden (Steen 1982). His observations are con-

CANE CHOLLA IS OFTEN ASSOCIATED WITH OLD RUINS. LONG HOUSE, BANDELIER NATIONAL MONUMENT.

DOVEWEED, AN INDICATOR SPECIES, GROWING AMONG THE RUINS AT FRIJOLES CANYON, BANDELIER NATIONAL MONUMENT.

sistent with those of other experts who have found a correlation of saltbush with prehistoric ruins; indeed, a site in Frijoles Canyon at Bandelier National Monument is officially named Saltbush Ruin in honor of the *Atriplex* that abundantly covers it.

Even with the extra nutrients and water available on some prehistoric sites, we still wonder how these plants came to be associated with ruins. No doubt an anthropogenic element is involved. Plants that tend to be concentrated on those ruins that have remained relatively undisturbed in recent times often had economic value for the prehistoric inhabitants.

The question is: Did the early inhabitants accidentally or intentionally introduce these plants to the site or was soil disturbance caused by human activity sufficient to create optimum conditions for the plants to become established on their own? Both ideas have merit, depending on the nature of the site and the plant species.

BUFFALO
GOURD
pg. 214

JIMSONWEED
pg. 204

HORSE-
NETTLE
pg. 208

In either case, in addition to the shrubs mentioned previously a number of other plants seem to indicate prehistoric disturbance. In his study of distinctive flora on Indian ruins at Bandelier National Monument, Richard Yarnell (Yarnell 1965) included the following herbaceous plants as "likely to be significant" indicator species used during occupation: beeplant, doveweed (*Croton texensis*), buffalo gourd (*Cucurbita foetidissima*), jimsonweed (*Datura meteloides*), groundcherry, horse-nettle (*Solanum elaeagnifolium*), wild potato (*Solanum jamesii*), and another species of Indian tea (*Thelesperma filifolium*). Nearly all these can be seen today growing on or near the various ruins in Frijoles Canyon, and some of them are found at the pueblo ruins of Coronado and Jemez state monuments. Note that these are all wild plants that happen to favor sites with past soil disturbance of some kind, whether from human-made or natural causes. Such disturbance did not necessarily take place in an archaeological context, and lists such as Yarnell's are of marginal value in locating buried sites.

Where ruins have been excavated or disturbed in the past fifty years or so, they are likely to be covered with exotic weedy annual plants, especially cheatgrass (*Bromus tectorum*), nonnative species of tansy mustard, summer cypress (*Kochia scoparia*), or Russian thistle (*Salsola kali*). These invaders of freshly disturbed land are especially common at Coronado and Jemez, as well as at Pecos National Historical Park, all sites with ruins that have experienced recent excavations.

Late historic agriculture was a notorious disturber of native plant habitats. It is less certain if the changes to natural vegetation patterns caused by late prehistoric and early historic agricultural practices would have lasted until the present day. Nevertheless, distinctive patterns of vegetation at sites believed to be long-abandoned fields or gardens have consistently been observed and recorded in the technical literature.

In the homestead claims reviewed for the Pajarito Plateau, nowhere is it mentioned that the claimants had to clear the land; they are only described as having had to break ground for their crops. In fact, the homesteaders of the late 1890s often sought out and plowed the sites of prehistoric field clearings surrounded by stands of oak, piñon, and juniper on the mesas. For the most part, the homesteaders, like those who came before them, grew corn and beans. At the edge of these homesteads we can still see the characteristic rectangular or L-shaped stone borders of prehistoric soil- and water-catchment gardens.

In his report of 1847, Lt. James W. Abert mentioned that a Spanish schoolteacher near Jemez told him the Indians grew beans on the mesa tops. On a never-plowed nearby mesa top we noted the same type of rock outline covering the surface. Whereas the remaining abandoned historic fields at Jemez are covered with introduced weeds, the prehistoric grid gardens have reverted to strictly natural vegetation. Small natural and human-made rock catchment basins are evident on these mesas, and they probably collected the water that would have been carried to the garden plots. On the other hand, in similar mesa-top locations the large plowed fields of historic times were directly dependent on rainfall.

Carl White, a biology professor at the University of New Mexico, has undertaken studies of prehistoric grid gardens near Ojo Caliente and along the Rio Chama. Grid gardens at these locations are usually outlined with large cobbles and covered with a smaller cobble or pebble mulch. Dr. White has found many of these gardens intact in sandy alluvium on top of steep hills, outwashed terraces, and knolls. The rock outlines and pebble mulch have stabilized the soil, retarded direct evaporation, increased retention of infiltrating water, and provided a refugium for certain plant species. For example, three species of grama grass, including black grama (*Bouteloua eriopoda*), are more abundant within the protected ancient gardens than on the surrounding land. These grasses have roots vulnerable to plant infection caused by cutting from sharp hooves. Only in those areas whose precarious location and grid garden walls have deterred domestic animals and wildlife do these grasses seem to thrive. Dr. White asserts that, because of this phenomenon, long-abandoned grid gardens probably would be identifiable on infrared aerial photographs, thus providing us with yet another plant indicator technique.

The broad valley bottoms of the Rio Grande and major streams surely were used for gardens or fields in the distant past; however, in most cases the evidence has been obliterated by recent farming, animal impoundments, and other developments. But in some of the less-disturbed drainages of the Pajarito, an unusual abundance of plants such as three-leaf sumac suggests that modern vegetation may still be influenced by agricultural practices of the distant past. In the southwestern part of our state, prehistoric terraces have been found with the progeny of an ancient cultivated agave crop still growing on them (Tierney 1973). Other plants indicate the arrival of the Spanish and still others immigrants from different countries, but these stories are for another book.

Suggested Reading

Potter, Loren D. and Richard Young
1983 Indicator Plants and Archaeological Sites, Chaco Canyon National Monument. COAS: *New Mexico Archaeology and History* 1(4):19-37.

Winter, Joseph C. and William J. Litzinger
1976 Floral Indicators of Farm Fields. In *Hovenweep 1975*, ed. by Joseph C. Winter. Archeological Report No. 2, Department of Anthropology, San Jose State Univ., CA.

Yarnell, Richard
1965 Implications of Distinctive Flora on Indian Ruins. *American Anthropologist* 67(3):662-674.

The Plants

TREES

Piñon pine 95
Ponderosa pine 99
Douglas-fir, White fir 102
Juniper 105
Willow 108
Cottonwood, Aspen 111
Gambel oak 114
Chokecherry 117
New Mexico locust 120

SHRUBS

Joint-fir 122
Yucca 124
Beargrass 127
Fourwing saltbush 129
Currant, Gooseberry 132
Apache plume 134
Mountain-mahogany 136
Threeleaf sumac 138
Cane cholla 140
Wolfberry 143
Broom snakeweed 145
Rabbitbrush 148
Sagebrush 150

GRASSES AND GRASSLIKE PLANTS

Broad-leaved cattail 153
Indian ricegrass 155
Common reed 158
Tall grasses 160

HERBACEOUS PLANTS

Nodding onion 163
False Solomon's seal 165
Wild buckwheat 167
Dock 169
Goosefoot 171
Amaranth 173
Four-o'clock 176
Common purslane 178
Western tansy mustard 180
Rocky Mountain beeplant 182
Doveweed 185
Globe-mallow 187
Cactus 189
Wild parsley 192
Milkweed 196
Blue trumpets 198
Scorpionweed 200
Bee-balm 202
Jimsonweed 204
Groundcherry 206
Horse-nettle 208
Wild potato 210
Paintbrush 212
Buffalo gourd 214
Cocklebur 217
Gumweed 219
Yarrow 221
Indian tea 223
Sunflower 225

6
PLANTS AND PLANTCRAFT

WELL OVER A THOUSAND different species of wild plants grow in the New Mexico Pueblo Province, that is, the area of the Rio Grande Valley from Taos to Isleta, the mountain ranges on either side of the valley, the Rio Puerco and San José river drainages west of the Rio Grande, and moving just over the continental divide to include Zuni territory. Nearly 300 of these plants are known to have been used in one way or another by contemporary Pueblo Indians or their ancestors, and the following sections cover about a quarter of them—those considered among the most important to the Puebloans and that may easily be seen in the wild today, especially along the trails of the four parks mentioned below.

We have grouped the plants by growth form—trees, shrubs, grasslike plants, and herbaceous plants—then by plant family in

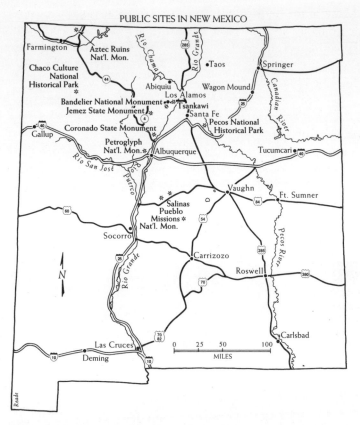

phylogenetic order, that is, those most closely related, which is the sequence found in most technical plant manuals. Each description includes the common and scientific name of the species or group of species. Sp. refers to singular species; spp. to plural species. In choosing nomenclature, we gave preference to the names used in the 1980 edition of *A Flora of New Mexico*, by Martin and Hutchins, but we borrowed some common names from other established sources when they seemed more appropriate for our area.

The principal distinguishing features are described for each plant, accompanied by a line drawing and, for most, a color photograph of the plant in its entirety. In addition, information on habitat, flowering season, and where to look for that particular species is provided.

Many of the herbaceous plants described can only be positively identified without question when they are in flower. Fortunately, the flowering period for most extends over many weeks and coincides with the season you are likely to be visiting the parks.

In some cases—for example, the pad-leafed prickly pear cacti—we deal with groupings of plant species difficult to distinguish without a technical plant identification key. Typically, the Indian users of such closely related plants have looked upon them as a single group, and so it is usually unnecessary to distinguish one species from the other in relating accounts of their use. However, when appropriate, we have included the current scientific names of the species in such groups.

In the descriptions we refer to four different parks where the plants can be seen and identified from public trails. Visitor centers located at the headquarters of each of the parks provide maps and other informational materials on the various trails. A brief description of the parks follows, along with addresses to write to for more information.

Bandelier National Monument About an hour's drive from Santa Fe, this park has a splendid network of more than seventy miles of back-country trails leading through spectacular scenery and passing numerous archaeological sites. You are likely to find the described plants on the Ruins, Frey, Falls, Upper Crossing, and Tsankawi trails, each covering a somewhat different habitat. Several of the trails have self-guiding interpretive booklets and markers that describe the plants and their prehistoric uses. A large bookstore and museum are in the visitor center, open daily all year; entrance fee for vehicles. Superintendent, Bandelier National Monument, Los Alamos, New Mexico 87544-9901.

Petroglyph National Monument, on the west side of Albuquerque, within the city limits, is one of the newer venues in the National Park system. It features thousands of petroglyphs carved into the black volcanic rocks. Many of these can be seen from the three short trails in Boca Negra Canyon covered in this book. Open daily; small entrance fee for use of trails. National Park Service, 123 Fourth Street SW, Room 101, Albuquerque, New Mexico 87102.

TOP
FRIJOLES CANYON, BANDELIER NATIONAL MONUMENT.

RIGHT
UPPER BOCA NEGRA CANYON, PETROGLYPH NATIONAL MONUMENT

Coronado State Monument, on the banks of the Rio Grande ten miles north of Albuquerque, preserves portions of the ruins from the Puebloan village of Kuaua, which was occupied from about A.D. 1325 through the early 1600s. Ancient kiva murals on display are considered among the finest examples of mural art dating from pre-European contact in North America. Many plants described on the pages that follow can be seen from the short trails here. Open daily; small fee for museum and trail through ruins. Box 95, Bernalillo, New Mexico 87004.

Jemez State Monument is an hour's drive north of Albuquerque on State Highway 4, west of Coronado State Monument. Ruins of a Spanish mission that served several pueblos occupied by Jemez people in the 1600s are the main attraction. The short trail to these ruins passes a number of the useful plants described in this book, and the small museum covers this subject from a Pueblo Indian point of view. Open daily; small fee for museum and trail. Box 143, Jemez Springs, New Mexico 87025.

Remember that the rules of all these parks (as well as those described in chapter 9, "Other Places to Visit") prohibit the disturbance of any natural or cultural feature. When it comes to the wild plants we cover or any other plants in the park, photograph and admire them but *don't pick them!*

The pueblos themselves are definitely *not* the place to go and look for these plants. In the first place, few pueblos have trails leading through natural areas open to the public. More important, Puebloan land should be thought of as private land. Its owners do not welcome uninvited guests wandering about their property any more than you or I would at home.

The descriptions of plant uses include both prehistoric and recent associations, if known. All references to Puebloan or prehistoric plant uses in this book come from original sources, either published materials, our own observations, or personal discussions that we had with Puebloans who gave us permission to include information on nonritual use in our book.

Some plants have so many uses that only their major ones can be described. We tried to cite specific pueblos wherever possible

RIGHT
THE RESTORED RUINS OF JEMEZ PUEBLO SHOWING THE MISSION CHURCH OF SAN JOSE, NEW MEXICO STATE MONUMENT.
Photo by Blair Clark

BELOW
KUAUA RUINS, CORONADO STATE MONUMENT

when discussing historic or current uses, but in some cases our citation is more general. For example, several of the early ethnobotanists wrote broadly of the Tewa Indians rather than citing one or more of the pueblos that comprise their language group, namely the San Juan, Santa Clara, San Ildefonso, Nambé, Pojoaque, and Tesuque pueblos. We must follow suit whenever generalized information is our only reference for a particular usage.

We would have liked to include the Indian names for the plants but found it impractical, since five distinct Puebloan languages (Tiwa, Tewa, Keres, Towa, and Zuni) are spoken in New Mexico, and none of these are written. But it is our special wish that Puebloans will nevertheless have use for this book as their own, writing in phonetically their people's name for a familiar plant, adding their own information, and excusing any misinterpretation we may have made.

The Concept of Ecozones

IN THE PLANT DESCRIPTIONS THAT FOLLOW, we often refer to the vegetative community, or ecozone, where that particular species is most likely to be found. Ecozones are defined by the dominant plants, usually one or more tree species, that characterize the pattern of vegetation in a given area.

Each zone is a reflection of the local climate, mainly the annual precipitation coupled with the average temperature, which, in turn, influences how much moisture is available for plant growth. The chart shown here depicts how in our area annual precipitation increases with elevation, resulting in a gradient of increasing annual moisture from the lowest land, along the Rio Grande, to the highest, near the summits of the Jemez, Sangre de Cristo, and Sandia mountains.

The effect of exposure (determined by the direction a slope faces) creates an additional influence on available moisture. South-facing slopes benefit from more direct sunlight and have higher daily temperatures throughout the year, whereas the colder, north-facing slopes retain soil moisture for a longer period. The result is that a given ecozone occurring on the north side of a mountain will be found at a considerably higher elevation—as much as a thousand feet higher—on the south side.

As you ascend from the Rio Grande Valley toward the mountains, you will notice that the patterns of vegetation reflect the increasing moisture and decreasing temperature. Desertlike vegetation of scattered shrubs and dry-looking grasses gives way to shrubby woodlands as you pass upward into the piñon-juniper ecozone that occupies much of the Pajarito Plateau. At higher elevation a taller evergreen forest is dominated by ponderosa pine, and, higher still, the even-denser forest, known as the mixed conifer ecozone, includes several coniferous tree species.

All the plants described in the following sections occur in one of four ecozones—juniper-grassland, piñon-juniper, ponderosa pine, or mixed conifer—although a fifth ecozone, spruce-fir, occurs at the top of the taller mountain ranges (chart). A nearby source of water increases the availability of moisture to the soil, of course, so

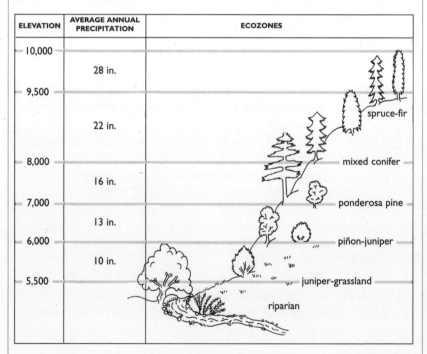

ELEVATION	AVERAGE ANNUAL PRECIPITATION	ECOZONES
10,000		
	28 in.	
9,500		spruce-fir
	22 in.	
8,000		mixed conifer
	16 in.	
7,000		ponderosa pine
	13 in.	
6,000		piñon-juniper
	10 in.	
5,500		juniper-grassland
		riparian

RIPARIAN ECOZONE.

a sixth vegetative community, the riparian ecozone, is manifested as a ribbon of water-loving trees and plants along the Rio Grande, El Rito de los Frijoles in Frijoles Canyon, and other perennial streams in New Mexico.

TREES

PIÑON PINE
Pine family
(*Pinus edulis*)

          *P*rehistoric Indians were the custodians of a once vast piñon and juniper "orchard" that provided them with materials for food, fuel, building, tools, and medicine. This woodland still survives, and it can be seen as you travel through the foothills on the way to Bandelier and Jemez monuments and toward most mountain ranges in New Mexico. The piñon tree is easily recognized by its rounded crown

PINUS EDULIS

and irregular shape. Needles are short and clustered in twos, and the bark is dark and rough. Open cones look like woody flowers and may persist on the tree even while immature cones are developing.

The great piñon woodlands where many prehistoric villages were located surely provided basic nourishment for the people prior to their cultivation of corn. Tewa Pueblo tradition still holds piñon nuts to have been their most ancient food. At Saltbush Ruin in Frijoles Canyon at Bandelier, whole piñon cones have been found in house ruins from the twelfth century, suggesting that, of the many ways of collecting piñon nuts, a simple one favored early on was to unscrew the pitch-covered cones from the branches and carry them home to open by heating or drying. This method of collecting early in the season would have avoided competition with piñon jays and other birds and animals. Once the cones opened and the twenty or so nuts in each cone were picked out, the empty, dry, pitchy cones would make good kindling.

Few trees produce prodigious seed crops every year, and the piñon is no exception. Flower formation is dependent upon moisture in late winter or early spring, and the cones and nuts do not mature until late summer a year later. The nut is fully developed, with all its considerable nutrient value, in late August. A bumper crop can be expected only about every six years. But surely in the past, as now, the Indians were still able to find some canyon or mesa where at least a few nuts could be collected. Back then there was no wholesale eradication of piñon trees on public and reservation lands in order to increase grasslands, no demand for charcoal to smelt silver and gold, no urban populations demanding firewood cut to size. But many believe that after the arrival of the Spanish the metal ax in the hands of Native Americans was indirectly responsible for denuding lands around ancient villages and pushing woodlands further away.

Containing over three thousand calories per pound, piñon nuts must have constituted the most valuable local wild plant food source for many prehistoric peoples living in the piñon-juniper ecozone. Moreover, the biological value of its protein on a per-pound basis is comparable to that of beefsteak and exceeds that of all commercial nuts, with the exception of the cashew. The piñon also contains all twenty amino acids that make up complete protein, and of the nine amino acids essential to human growth, seven are more concentrated in piñon nuts than in corn.

In recent times Indians from all pueblos have collected piñon nuts, setting up temporary camps in the woodlands when seeding begins and beating or shaking the nuts onto blankets or canvas. The nuts would be toasted on a stone griddle at the site, then carried home in jars or sacks, where they would keep for several years. In a good year a surplus could be easily acquired for home use, trade, and, in modern times, a cash crop.

Piñon nuts are the most sought after of wild plant food by contemporary Pueblos as well as many other New Mexicans. At San Juan Pueblo it is the only plant product that family members still gather as a household activity. Once the irrigation ditches are closed in the fall, many Pueblo families will merrily spend a week collecting nuts in the woodland hills surrounding their village. It has been reported that, raking up the litter beneath piñon trees and picking up the nuts immediately after seed fall, a husband-and-wife team could harvest 150 pounds of nuts in a single day. Half of this harvest is shell, leaving about 70 pounds of nut meat, or the equivalent of more food energy than the same weight of chocolate—but without the saturated fat.

The miscellaneous historical uses of the piñon are legion, but we include only a few here. At Jemez, in the early part of the twentieth century, a red pottery paint was obtained by mixing the resin of old and new trees. Also at Jemez, a blue-green or turquoise paint was fabricated by boiling piñon gum. An all-purpose glue could be made by warming the pitch, and this was used to secure turquoise in jewelry settings and sinew to the back of bows. With the addition of a wrap of wet sinew, piñon glue was also used to secure stone points and feathers to arrows.

Piñon pitch is still used medicinally by Rio Grande Puebloans, powdered and sprinkled in wounds as an antiseptic or mixed with warm tallow or candle wax and placed on sores to draw out infection. It is also chewed as a gum or chewed and swallowed for clearing the head during a cold.

Felipe Lauriano of Sandia Pueblo recounts how piñon gum was collected, especially from trees where porcupines had chewed the trunk in winter, causing pitch to form. His grandfather said he used the gum on the rawhide soles of moccasins to keep them flexible and maintain their shape. It was also dabbed on arrows to strengthen the end where the feathers and the notch for the bowstring went and, in combination with certain flowers, to color this end of the arrow. Each pueblo was said to have used a different color and shape of feather clipping.

PONDEROSA PINE
Pine family
(*Pinus ponderosa*)

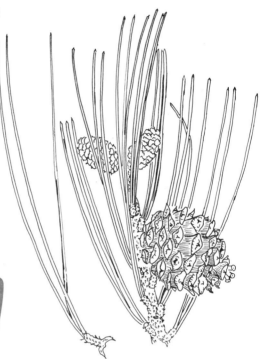

Named for its ponderous size by a botanical explorer in 1826, this is the largest native evergreen tree in our area. Often growing to a height of more than 125 feet, it can live to be well over 300 years old under favorable conditions. Because of the longevity of ponderosa and its sensitivity to seasons and drought, trunk rings of this tree are especially useful in determining prehistoric chronology and climate (see chapter 7, "Ethnobotany in New Mexico").

The mature ponderosa pine is irregularly pyramidal in shape, but the lower half of the trunk tends to be free of limbs. Higher branches frequently turn upward toward the outer ends. Needles are five to ten inches long and come in bundles of three, with a tiny papery sheath surrounding the base of each bundle. The bark of immature trees is black and rough but that of mature trees is deeply fissured into diamondlike patterns of red or yellow-brown platelets. The bark sometimes has the scent of vanilla.

Ponderosa is the most commercially valuable pine in the

PINUS PONDEROSA

Southwest. Both young and very old ponderosa pines are common throughout Bandelier National Monument, mostly between 6,000 and 8,000 feet elevation, although the shaded bottom of Frijoles Canyon harbors many fine specimens.

In times of need Puebloans and other groups of Indians in the Southwest would chew the soft inner bark, or cambium, of the ponderosa for its nutritional value. The cambium can be very bitter, so if used in any quantity, it was probably processed by pounding or grinding and then leaching with water. When faced with starvation, people around the world have survived by using their traditional knowledge to extend their food supplies with the cambium and leaves of trees.

At San Juan Pueblo, ponderosa needles have been chewed as a cold sore remedy, and a *decoction*, or boiled concentrate, of the root used to be drunk for urinary problems. The needles, which contain vitamin C, make a pleasant tea. On a cautionary note, some cattle abortions have been reported following their consumption of pine needles. It would seem, though, that only the lack of forage would force animals to use pine needles as food.

At prehistoric Bandelier, ponderosa pine was the predominant wood used for roofing supports, especially for the main cross beams, called *vigas*. It was preferred whenever straight-grained wood was desired for manufacturing such things as ladders or the backs of cradleboards. At one Anasazi site in Frijoles Canyon, "two pieces of yellow pine were used to fashion the board for use in a cradle board. They are both approximately 14 inches long, 4 inches wide and 11/16 inch thick....They have been worn quite smooth, undoubtedly, through use" (Turney 1948).

Cradleboards have been and still are routinely used by Rio Grande Puebloans, with the family board passed on from one generation to the next. It wasn't until after World War II, when soldiers from the Pueblos returned home with their crew cuts, that one of the effects of this tradition was noticed. Hard cradleboards can deform a baby's soft skull by flattening the back of the head and thereby broadening the face. Some Indian tribes escaped this deformation of the head because they traditionally employed soft, wicker-backed cradle "boards" for their infants.

DOUGLAS-FIR
Pine family
(*Pseudotsuga menziesii*)

WHITE FIR
(*Abies concolor*)

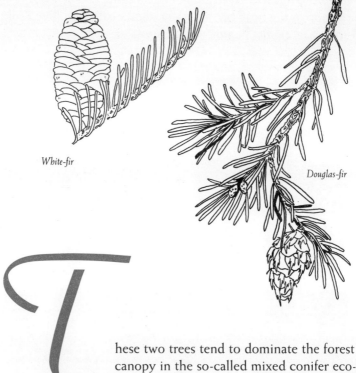

White-fir

Douglas-fir

These two trees tend to dominate the forest canopy in the so-called mixed conifer ecozone, which occurs mostly above 7,500 feet elevation in our area. Both have short, flat needles and somewhat of a Christmas-tree growth form. Douglas-fir (not a true fir) is the larger of the two, growing up to two hundred feet tall and bearing cones that hang down from the branches and have papery bracts protruding from between the scales. White fir reaches only about one hundred feet tall, has a distinctly bluish cast to the foliage, and has no obvious bracts in its upright cones.

At Bandelier both of these trees are fairly common along the Upper Crossing Trail and even more so on the Ski Trail at the upper

ABIES CONCOLOR　　　　　　**PSEUDOTSUGA MENZIESII**

end of Frijoles Canyon. Douglas-fir also descends into the lower, wetter drainages and can be seen growing throughout the riparian area near the park visitor center. Neither tree is found on the trails of the other parks covered in this book because the elevation is too low there.

Although the two trees are similar in appearance, Puebloans know them as separate species. Douglas-fir enjoys more significance by far; in fact, it is probably the single most important wild plant associated with Pueblo ceremonies. Thus, it's not surprising that objects appearing to be prayer sticks made of this wood have been excavated from several Anasazi-era sites.

All the pueblos seem to have one or more dances that incorporate Douglas-fir into the ceremony. The twigs may be carried or worn on various parts of the dancers' apparel, as at Zuni, or the wood may be incorporated in masks or other parts of the costumes. One original Kuaua Pueblo mural dating to about A.D. 1450, and now reproduced in the main kiva (ceremonial chamber) at Coronado State Monument, shows a figure holding a stylized tree with needles, surely a depiction of the sacred Douglas-fir.

Douglas-fir has very few medicinal applications that we know of, although Isleta Puebloans apparently used to steep a tea from the needles for curing rheumatism. As a wood for vigas and other parts of ancient dwellings, Douglas-fir ranks a distant second to ponderosa pine, at least among the prehistoric sites on the Pajarito Plateau. However, limbs from this tree were used as a roofing material at San Juan Pueblo as recently as the late 1960s.

Although white fir appears to have no ritual connections, several medicinal uses have been attributed to it. These include a hot tea from the foliage to combat rheumatism at Acoma and the use of resin for cuts by the Tewa-speaking Puebloans and for relief from earaches by the Picuris Indians. In old times various Pueblo people used white fir bark in a solution mixed with wild dock (*Rumex* spp.) and joint-fir (*Ephedra* spp.) to tan deer skins.

ONE-SEED JUNIPER
Cypress family
(*Juniperus monosperma*)

ROCKY MOUNTAIN JUNIPER
(*Juniperus scopulorum*)

ALLIGATOR JUNIPER
(*Juniperus deppeana*)

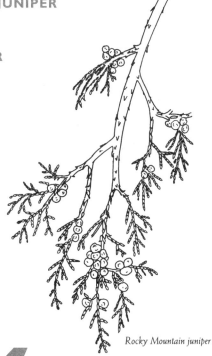

Rocky Mountain juniper

Most species of juniper in New Mexico take the form of large shrubs or small trees. They can be recognized by their tiny scalelike aromatic leaves and hard, bluish, pea-size berries. In our area one-seed juniper can be identified by its shreddy bark and the single large seed encased in its berry; Rocky Mountain juniper berries usually contain two seeds; alligator juniper has the characteristic checkered bark that gave it its name.

JUNIPERUS MONOSPERMA

Along with piñon pine, one-seed juniper is the dominant tree throughout the lower reaches of the middle Rio Grande watershed. It is common along all the lower trails at Bandelier as well as at Jemez State Monument; at Petroglyph National Monument only occasional trees are seen. Rocky Mountain juniper favors wetter conditions and occurs along Frijoles Creek as well as at higher elevations, such as on the Upper Crossing Trail. Alligator juniper is rarely encountered on the trails covered in this book.

Juniper has had enormous economic importance to all Pueblo Indians and still enjoys many uses today. In fact, a greater variety of uses, including food, medicine, construction, and crafts, may be attributed to juniper than to any other group of plants.

Junipers' annual berries formed a food staple in ancestral Puebloan diets. Indeed, because of the reliability of this crop, even during periods of drought, the berries are thought to have been one of several famine foods in times past. Juniper berries, eaten raw or stewed, continued to be regularly harvested by Rio Grande Pueblo Indians up until the supermarket era. Today, the berries are more likely to be used for seasoning meat and stews than as a food staple.

A tea made from juniper berries has worked as a diuretic and for internal chills by Santa Clarans and tea from leaf sprigs for colds, stomach disorders, constipation, and rheumatism by various other Puebloans. A weak juniper tea is said to act as a stimulant, a strong one an emetic.

Among Puebloans there has been an almost universal association between juniper and the process of giving birth. Virtually all Pueblo women have a tradition of drinking juniper sprig tea either during labor or immediately after a child is born, and many have used a liquid concentration for bathing both mother and baby at that time.

Juniper bark has been boiled and bathed in by Sandians to relieve itch from spider bites, and a powder from inside the bark has been used by Cochitis for earaches. The inhabitants of several pueblos burn juniper branches in their homes when fumigation is needed, sometimes to relieve colds, or often simply for the clean, woodsy fragrance.

The wood has been used in endless ways, including for fuel, construction, and smaller implements, such as bows, digging sticks, and basket frames. The soft bark was formerly employed in matting, and in the late 1940s the young women of San Ildefonso were still crushing and shredding juniper bark as an absorbent packing around the bottoms of their babies to keep them dry and sweet smelling.

Several artisans at San Juan, a pueblo noted for its lovely plant-formed jewelry, work drilled and shellacked juniper berries into their handcrafted seed necklaces. The San Juan Pueblo Crafts Cooperative displays and sells a fine collection of this jewelry.

We can't end the discussion of juniper without mentioning juniper mistletoe (*Phoradendron juniperinum*), a yellow-orange plant parasite whose dense clusters often cling to the outer limbs of juniper trees. Like the tree upon which it lives, juniper mistletoe also produces berries, translucent globules that sometimes supplemented the diets of the ancients when other food became scarce. These berries, too, have had medicinal applications, including the prevention of baldness (at Santa Clara), a cure for diarrhea (for children at Acoma and Laguna), and an emetic for stomachache (at Zuni and the various Tewa-speaking pueblos).

WILLOW
Willow family
Coyote willow (*Salix exigua*)

Several species of willows grow along the banks of rivers and streams in our area. All are shrubs or small trees with characteristic narrow leaves, but the commonest, coyote willow, can be recognized by its linear leaves with a silvery undersurface. Coyote willow grows in thickets along the Rio Grande and lower-elevation feeder streams such as El Rito de los Frijoles at Bandelier National Monument.

Throughout Indian country in the Southwest, slender branches of coyote and other kinds of willow have been and still are the preferred material for making circular coiled baskets and trays. New spring growth works best, as it is more pliable. One well-preserved wicker basket recovered from an Anasazi ruin in Frijoles Canyon, near the visitor center at Bandelier, was described as follows:

SALIX EXIGUA

The maker of this basket first obtained sixteen twigs approximately $1/8''$ in diameter. Four sets containing four each were arranged overlapping each other in an over and under process. They were then bound with yucca cordage. After the center was started in this manner, the forty radiating groups of twigs were arranged like the spokes of a wheel.... Many of these were inserted as the basket progressed so that the basket would have larger spokes at the center than at the bottom or at the rim. When the desired circumference had been reached, the projecting ends of the ribs were bent sideways and two spokes merged into one at the rim. A bundle of willow twigs encircles the top and is fastened to the basket proper with strands of yucca leaves. This formed the rim of the basket (Turney 1948).

Other implements made from willow include fire sticks twirled as a spindle to generate enough heat to ignite a flame and what appear to be prayer sticks recovered from various archaeological sites. Willow is still used for making prayer sticks by the Zunis and doubtless by some of the Rio Grande pueblos.

Coyote willow branches are much too thin for use in heavy construction, but the straight limbs, sometimes with leaves attached, used to be employed at various pueblos for thatching roofs. Other light construction uses would have included the tops of storage bins or racks for aerating corn while it dried, such as one recently unearthed at prehistoric Arroyo Hondo Pueblo.

Willow bark and stems are an important source for salicin, which, when ingested, breaks down into salicylic acid, the basic ingredient of aspirin. Long before aspirin was being championed as a universal pain reliever, Pueblo Indians had learned to brew willow bark tea to relieve coughs and sore throats or for a soothing skin bath. Zuni Puebloans also used to chew the bark to relieve sore throats and toothaches. At San Juan a tribal elder told us he used to munch willow leaves to make his mouth water and relieve thirst. We tried it too, and it works!

FREMONT COTTONWOOD
Willow family
(*Populus fremontii*)

NARROWLEAF COTTONWOOD
(*Populus angustifolia*)

ASPEN
(*Populus tremuloides*)

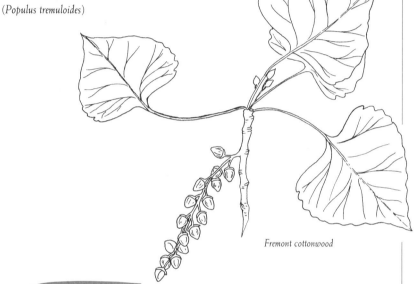

Fremont cottonwood

The two kinds of cottonwood that inhabit our area never occur far from watercourses. A third, closely related tree, the aspen, grows in the high mountains. All three are deciduous, their foliage brightening the landscape with yellow to golden hues in fall before dropping for the winter.

As described in chapter 8, "Recent Modifications to the Landscape," there aren't nearly as many Fremont cottonwood trees in the bosques of the Rio Grande as there once were, but they still are

POPULUS FREMONTII

the dominant tree along many sections of the river, such as at Coronado State Monument. Narrowleaf cottonwood, with its willowlike leaves, is the cottonwood of mountain streams and dominates the riparian canopy in Frijoles Canyon at Bandelier. Aspen, with its characteristic smooth white trunk, is abundant above 8,000 feet elevation, growing in patches or sometimes on entire mountainsides, particularly where it has colonized burned areas.

Cottonwood and aspen are known as *the* woods for making drums. The wood of other large-trunked trees in our area—ponderosa pine, Douglas-fir, and white fir, to say nothing of piñon or juniper—is either too gnarly or too heavy or it lacks good resonating qualities. But most important, cottonwood and aspen trunks rot in the center after the trees die, making the hollowing process easier. At one time every pueblo would have counted drum makers among its artisans. Today Cochiti, Taos, and Jemez pueblos are best known for this craft, although drum making is also pursued at a few other modern pueblos.

Aspen is most often used for small- to medium-size drums; narrowleaf cottonwood, when available, is preferred for larger ones. The wood of all three species is straight grained, light, and relatively easy to carve, but it takes even the experts many days to produce a single drum.

Once most of the Indians moved off the plateaus and down to the Rio Grande drainage hundreds of years ago, Fremont cottonwood undoubtedly became the most useful broadleaf tree for all sorts of construction, including the making of small boats and rafts by Isletans and probably other Puebloans who lived along the river.

POPULUS TREMULOIDES

The dense but soft roots of cottonwoods are one of several materials used to carve kachinas.

Drooping cottonwood flowers, called *catkins*, appear on the trees in spring before they leaf out. Catkins once were eaten raw at various pueblos, including Jemez, Isleta, and San Juan, and at San Ildefonso the catkins were an early-spring vegetable ingredient of meat stews.

All these trees contain chemical substances that have pain-relieving qualities similar to those of aspirin. Thus, at San Juan Pueblo, among others, cottonwood leaves once were applied to aching teeth and to skin abrasions. At San Ildefonso the leaves were boiled to make a tea for urinary problems. One unexpected medical application results from the ease with which tough Fremont cottonwood bark can be stripped in chunks from the trunk. At more than one pueblo cylinders of cottonwood bark were once used for making bone splints.

The wood of these trees burns hot and clean. As related to us by a potter we met at the Tesuque Pueblo Senior Center, this is the favorite wood for firing pots. Cottonwood bark also is known to have been used for this purpose.

GAMBEL OAK
Oak family
(*Quercus gambelii*)

Several species of oak grow in the Rio Grande Pueblo Province. They range in size from low shrubs to medium-size trees, and all bear acorns—those recognizable small nuts partly surrounded by a scaly cap. The various species frequently hybridize, the results of which can be a bit confusing to amateurs and botanists alike. Here, we focus on Gambel oak, which has characteristic deeply lobed leaves, unlike the other oaks found in our area.

Gambel oak typically grows in dense thickets throughout the piñon-juniper woodlands upward into the ponderosa pine ecozone, often taking over the ground surface for decades following a severe wildfire. If untouched by humans or fire, this species can attain the size of a huge tree when isolated plants become established on the rich soil of valley bottoms. At Bandelier you will pass through

QUERCUS GAMBELII

thickets of Gambel oak below the Ponderosa Campground on the Upper Crossing Trail, which skirts the edge of the La Mesa Fire of 1977. Gambel oaks are also common near the stream at the bottom of Frijoles Canyon.

Naturally, being a common plant with edible nuts growing in the heart of occupied land, this species was heavily utilized from Archaic times right up to the present century. In fall the acorns were collected and ground into a protein-rich meal for mushes or cakes. Many species of oak, including some that once provided a basic food source for most Indians in California, bear acorns with a very high tannin content. Tannin makes them so bitter that the meal must be leached for hours in water before it is edible. Tannin-poor Gambel oak acorns, on the other hand, are actually somewhat sweet and can be eaten raw with no ill effects, as described to us by a man from Cochiti Pueblo who likes to chew on them whenever he's up in the mountains during deer season. At Santa Clara acorns may be cut into small pieces and used as a pie filler.

Lysine, an amino acid essential for human nutrition, is concentrated in oak acorns, as it is in beans. But lysine is lacking in corn, the mainstay of the people in these parts for hundreds of years.

Gambel oak acorns could, and probably did, provide an important complement to corn whenever the local bean crop was insufficient.

Gambel oak boughs, tough and pliable, were used by different Pueblo people for making trays. Stouter wood was employed for digging sticks, clubs, and tool handles and sometimes for bows or arrow shafts. Two weaving sticks made of oak were recovered from a ruin dating to about A.D. 1500, located in Frijoles Canyon near the Bandelier Visitor Center. Remains of oak wood are almost always present among the excavated ruins of such prehistoric villages.

Michael Moore lists countless applications for oak parts in his several books on medicinal uses of plants (such as Moore 1979), but recorded medicinal uses by Puebloans in our area are scanty. The relative lack of tannin and perhaps other active chemicals in Gambel oaks may account for this anomaly.

CHOKECHERRY

Rose family
(*Prunus virginiana*)

Usually a small tree, sometimes a shrub, the chokecherry has twigs and bark with lens-shaped reddish brown markings. In early summer its showy elongated clusters of small white flowers tend to hang downward, as do the ripe purple-black fruits later in the season. But you aren't likely to see many ripe fruits since birds and squirrels quickly gobble them up.

There are several species of wild cherries and wild plums along the watercourses and mountain slopes of New Mexico, but the chokecherry is the most common species. It can be seen on the walk to the Lower Falls at Bandelier and throughout the riparian zone in Frijoles Canyon.

The remains of chokecherry pits have been found in desiccated human feces (coprolites) at several Anasazi sites in northwestern

PRUNUS VIRGINIANA

New Mexico, in room fill at Arroyo Hondo Pueblo, and in charred material from the Pajarito archaeological site at Bandelier. These and other discoveries in prehistoric sites throughout the United States indicate that the wild cherry was continentally abundant then as it has been historically.

At Bandelier National Monument a land feature called Capulin Canyon (*capulin* means cherry in Spanish) indicates that these trees probably grew there in abundance at one time. Feral burros love young chokecherry twigs and will knock a tree down to get at them. Chokecherry destruction in the canyon, caused when these creatures ran rampant years ago, is so obvious that perhaps the place should be renamed *Sin* Capulin *("without* cherry").

The strong, supple, straight-grained cherry wood was used to make functional bows at the pueblos of Isleta, Acoma, Laguna, San Juan, and San Ildefonso until at least the first half of this century. Chokecherry bark and roots are used by American Indians in various traditional medicines, and the Puebloans are no exception. A cough medicine is made of the bark, and the ground root has been sprinkled into wounds. In the eastern United States, black chokecherry bark is collected and sold commercially as a proven remedy for the same maladies, and remedies constituted of these ingredients can be purchased at local health food stores.

The best-known use of cherry fruit is as a food. Known as "bitter hanging fruit" at Zuni, it was the basis for different sauces. The Lagunas travel to the slopes of Mount Taylor (northeast of Grants, New Mexico) to gather chokecherries, and the Tesuques collect them at Rio en Medio (a valley northeast of Santa Fe) during trips for firewood. Probably at all the pueblos, depending upon degree of availability, these somewhat astringent fruits are eaten fresh, stewed with sugar for desserts or preserves, or dried for later use.

As with cultivated fruit, drying wild cherries concentrates the sugar content. Drying also tends to mitigate the bitterness that might make wild fruit less edible. We have found this to be true of a variety of wild fruits not thought to be palatable before drying.

NEW MEXICO LOCUST

Pea family
(*Robinia neomexicana*)

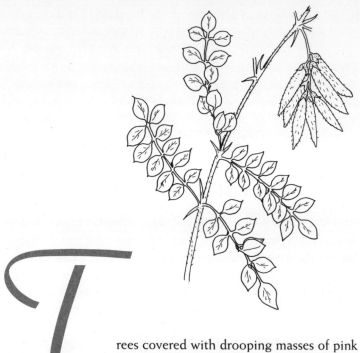

Trees covered with drooping masses of pink blooms are a common sight alongside many mountain roads in New Mexico in early summer. This species is the only wild locust tree in our area, and it can be identified throughout the growing season by the numerous pairs of elliptical leaflets attached to each leaf stem and by its short-spined twigs. New Mexico locusts seldom grow much taller than twenty-five feet, and sometimes, especially when in thickets, they look more like large shrubs.

These trees favor canyons or slopes from the upper piñon-juniper zone up into the ponderosa pine and mixed conifer ecozones. Look for them along the roads approaching Bandelier, at the bottom of Frijoles Canyon, and on the Upper Crossing Trail, where it passes along the edge of the La Mesa Fire of 1977. One of the best flowering locust displays in the state is on the road to Sandia Crest east of

ROBINIA NEOMEXICANA

Albuquerque where, in places, both sides of the pavement are lined with hues of rose-pink from late June through early July.

The branches of New Mexico locust are tough and elastic—ideal for making bows, as was done in historic times at several pueblos, including San Juan and Jemez. This tradition goes back at least to the Anasazi era, for one end of a bow made from this wood was recovered from a 500-year-old ruin not far from the main visitor center at Bandelier. Arrow shafts also were made from locust tree branches.

People from at least one pueblo, the Jemez, used to prize the uncooked flowers as an early-summer addition to their diet. The attraction, no doubt, was the sweetness of the nectar.

SHRUBS

JOINT-FIR
Joint-fir family
(*Ephedra* spp.)

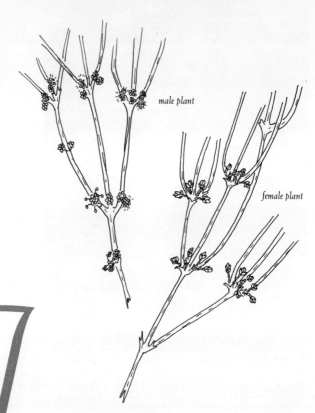

male plant

female plant

Joint-fir is a sparse-looking shrub, up to a yard or so tall, with jointed stems, scalelike leaves, and opposite or whorled branching. It favors open, dry rocky, or sandy places at lower elevations, mainly in the piñon-juniper ecozone or below. You will encounter this shrub along the Falls Trail at Bandelier as well as at Petroglyph.

Its best-known use is as a hot beverage, and the plant also is known as "Mormon tea." Joint-fir tea isn't favored as a day-to-day beverage for Puebloans nearly as much as the brew made from

EPHEDRA SP.

Indian tea plant (*Thelesperma megapotamicum*)—the former just doesn't seem to have as good a flavor. But it is still collected and brewed by people from various pueblos. It's also considered one of the best plants for quenching thirst in the field. One man told us that when he would ride all day checking fence lines in the foothills above Sandia Pueblo, he never needed water because, when thirsty, he just chewed on some stems from this plant and could work all day long.

Contrary to popular belief, joint-fir plants growing in the Southwest don't contain the chemical ephedrine, as do Old-World species of *Ephedra*. However, two other alkaloids, tannin and pseudoephedrine (the active ingredient in some commercial decongestants), are present in our species and may account for the various medicinal uses of joint-fir attributed to Puebloans. Tea brewed from leaves and stems has been drunk for urinary disorders at Santa Clara, Acoma, and Laguna and for diarrhea at several Tewa-speaking pueblos. The tea has also been used for cough medicine at Acoma and applied as a lotion for itchy skin at Isleta. Plant parts from joint-fir have been ingested for hundreds of years, as attested by coprolites analyzed at several sites dating from the Anasazi era.

Owing to the tannin in joint-fir, a powder ground from boiled plant parts has been employed during historic times by Puebloans to rawhide. The use of plant materials such as joint-fir, oak, and wild dock in tanning was learned from the Spanish. Previously, tanning was done using urine or animal brains.

BANANA YUCCA

Lily family
(*Yucca baccata*)

NARROWLEAF YUCCA

(*Yucca glauca*)

narrowleaf yucca

Both of these yucca types are widespread on the Pajarito Plateau and in the Rio Grande Valley, mainly on grasslands and rocky slopes in the piñon-juniper zone. They are easily recognized by their two- to three-foot-tall growth form—a dense cluster of blade-like, sharp-tipped leaves that usually bear gray or white fibers along the edge, all stemming from a central base at ground level.

The banana yucca is distinguished by its broad, stiff, succulent leaves, which are up to three inches wide. The leaves of narrowleaf yucca are flat, somewhat flexible, and less than an inch wide. All New Mexican yuccas produce creamy white flowers on a single erect stalk in late spring, followed by a series of large, fleshy green

YUCCA BACCATA

fruits that usually remain on the stalk until consumed by wildlife.

Banana yucca is encountered along several trails at Bandelier and Petroglyph national monuments. At Bandelier look for both types on the rocky sides of Frijoles Canyon and on the lower end of Frey Trail. Narrowleaf yucca is fairly common along the path to Tsankawi, whereas banana yucca tends to occur at slightly higher elevations, such as on the steeper slopes of the Upper Crossing Trail below Ponderosa Campground.

Yucca was used in many different ways and probably had greater economic importance to Pueblo Indians than any other group of wild plants growing in this region. Indeed, it is one of the few wild plants that occasionally was represented symbolically in prehistoric times, for example, on several of the petroglyphs in Boca Negra Canyon at Petroglyph National Monument.

The thick, sweet fruit of the banana yucca formed a staple of prehistoric diets throughout the region. The "bananas" would have been eaten green or dried and stored for winter use. At San Ildefonso the pulp was often mixed with chokecherries and made into a cake, and at various other Rio Grande pueblos ripe fruit was cooked into a paste and dried for future use. Young flower stalks of narrowleaf yucca were eaten occasionally. Usually baked in a slow oven, banana yucca fruit continues to be prized by latter-day Puebloans.

Fiber from the leaves of several different species of yucca played an even more important role in the lives of the early people. In fact, throughout the Southwest no material was in greater demand for manufacturing cordage than yucca leaf fiber. Yucca "quids" are commonly found in archaic cave sites. The flesh of the leaves was chewed off, leaving a fibrous mass for later use.

Historically, the long, straight leaves were soaked, then pounded with stones to extract the fibers, which were then twisted to fabricate string or rope. Sometimes human or animal hair was incorporated into more delicate strands that may have been intended for decorative or ceremonial purposes. Among documented applications of yucca cordage at Bandelier and surrounding sites are belts, rope ladders, toe cords and lashings for sandals, head rings for carrying water jars, and straps, cradle lashings, and fishnets, the latter used by Puebloans living on the Rio Grande.

The use of yucca fiber for matting and in basketry was and is widespread. Cloth once made from yucca cordage sometimes incorporated fur or feathers in the weave. One of the early excavations of the Frijoles ruins at Bandelier, for example, turned up a delicate blanket composed of rabbit skins combined with yucca fiber. Narrowleaf yucca is preferred by modern basket makers who wish to employ traditional methods.

For ages an extract made from saponin-rich yucca roots was the equivalent of soap for virtually all Indians living in the Southwest. The root of this plant is still regularly sought by Pueblo Indians for concocting a hair shampoo. The dry roots are pounded, then thrashed in cold water, the suds lathered into the hair and scalp. Yucca root shampoo is known to give black hair a shiny look prized by ceremonial dancers. Zuni men always wash with it before a dance, as do many women at Cochiti and Jemez pueblos. It has been said that washing hair with yucca shampoo strengthens the strands and even prevents baldness.

Before commercial implements were available, yucca fibers were used by women at many of the Rio Grande pueblos to make brushes for rendering designs on their pottery. The leaf was chewed until the tip was finely frayed, then the fringe was employed as a paintbrush. A number of contemporary Pueblo artists, notably several at Acoma Pueblo, still use a traditional narrowleaf yucca brush in decorating their beautiful handcrafted pots.

BEARGRASS

Lily family
(*Nolina microcarpa*)

With its spread of long, narrow leaves and dense cluster of white flowers on a tall stalk, beargrass resembles narrowleaf yucca. But this closely related species differs from yucca in that its leaves are much longer, triangular in cross section, with rough margins of minute teeth. The flowers are tiny, and the round fruits are only about a quarter-inch in diameter. Although beargrass is common on dry hills and mesas throughout the Pueblo Province, it has not yet been spotted along the trails covered in this book.

Earlier in this century beargrass seems to have been the preferred material for making baskets, especially among the more southerly Rio Grande pueblos. The process was described by two biology professors at the University of New Mexico:

NOLINA MICROCARPA

In the manufacture of the baskets the leaves were first plaited on the floor into a mat of the desired size. The mat was then fashioned into a shallow basket by tying the edges outward over a stick which was bent in the form of a circle. In the case of small containers the stick was usually removed although in the very large ones it often remained to serve as a rim. Such baskets were very coarsely constructed, and never decorated since they were not an article of trade but made with the idea of serving as inexpensive containers (Bell and Castetter 1941).

At Isleta Pueblo flour was once made from ground beargrass seeds, and the fruits also were eaten. But care must have been taken not to include the flower stalks in the meal, since these parts contain toxic saponins. This plant is highly poisonous to sheep and goats, though less so to cattle.

Also at Isleta drinking the liquid from boiled beargrass roots has been considered to be a remedy for pneumonia and rheumatism. Food or medicinal use of this plant has not been recorded for any of the other pueblos in New Mexico. However, at least one pueblo has used the dried seeds in dance rattles made from cultivated gourds.

FOURWING SALTBUSH
Goosefoot family
(*Atriplex canescens*)

Saltbush is an irregularly branched, sometimes spiny shrub with thick, gray-green, canoe-shaped leaves. This shrub is *dioecious;* that is, the male and female plant parts appear on separate plants. Male flowers are arranged in small *glomerules*—rounded dense clusters—on spikes along the branchlets; female flowers produce a tiny, oblong seed with four obvious paperlike wings.

Saltbush is undoubtedly one of the most valuable forage shrubs in arid regions of the Southwest. Able to exist on lands heavily impregnated with white alkali, it is found under almost any conditions in New Mexico, including in gravelly washes, on mesa tops, ridges, and hillsides. It's therefore common along virtually all the trails covered in this book.

Though saltbush does well on deep, sandy soils and occasionally grows on sand dunes, it's not considered to be an indicator of

ATRIPLEX CANESCENS

any particular set of substrate conditions. Yet it does seem to have an affinity for the prehistoric sites of the Pajarito Plateau, an affinity so marked that some scientists have deemed saltbush to be an indicator plant for ruins. Saltbush Ruin at Bandelier is an excellent place to view this phenomenon, as are slopes below the cliff dwellings throughout the Pajarito Plateau.

Like almost every common shrub in our area, saltbush has proven useful to native peoples past and present. Many years ago the nutritious seeds were ground and cooked as a cereal; the leaves of the plant were also eaten cooked and sometimes raw or were dried and mixed with other ingredients to form a flour for breads and cakes. Ashes of burned saltbush are used as leavening for breads, as a food coloring, and in the making of lye to soften the hulls of corn. All these uses are, incidentally, an effective way of making the niacin in the inner grain available, thus enhancing the amino acid content of potentially digestible proteins in corn. In societies that traditionally process maize with alkali in this way, there's little evidence of the dietary deficiency disease pellagra.

Many other uses of saltbush are known. The ashes of burned saltbush are still used in Hopi country to impart a greenish hue to the finely ground blue cornmeal of *piki* bread. Also, because of the ubiquity of the shrub, it's not surprising that saltbush has been used whole to caulk or cover the roofs of adobe dwellings. The shape of the main stems has also suggested many uses as tools and kitchen utensils. Having many tines protruding from the central stalk, a peeled branch from a browsed shrub makes a good whisk for stirring large pots of cereal or cornmeal mush; use of such an impromptu whisk has the dividend of imparting a mildly salty taste to the otherwise bland ingredients of a campfire meal.

The hard twig ends of saltbushes, tending to be spiny or pointed, were further shaped by carving and used for arrowheads inserted into the hollow ends of reedgrass or other shaft material. This type of point was used for the swift or war arrow and was considered poisonous or infectious by the people of Isleta. Arrows with stone points were used for game.

At Zuni handfuls of blossoms (probably male) are crushed and mixed with a little water and used like a hand soap. This suggests that fourwing saltbush contains saponins. Crushed flowers, either dried or fresh, are also used topically for ant bites, and the roots of another species of saltbush, *Atriplex argentea*, are used for skin sores and rashes.

CURRANT
Saxifrage family
(*Ribes inebrians* and similar species)

GOOSEBERRY
(*Ribes inerme* and similar species)

wax currant

The most common wild currants and gooseberries in our area are many-branched shrubs that seldom get as tall as human beings. Creamy pink or yellow bell-shaped flowers of late spring produce red or dark purple berries by the end of summer. Currant bushes are spineless, but gooseberries have spines or bristles on the stems where the leaves are attached.

Currants and gooseberries grow in a variety of habitats, including canyon bottoms such as at Frijoles, where both are plentiful along the park trails. Jemez State Monument is another place where gooseberries thrive.

RIBES INEBRIANS

Like most cultures in North America, all nineteen pueblos of modern New Mexico have a tradition of collecting and eating wild currants or gooseberries. The fruit may be consumed fresh or dried for cooking and snacks. Unfortunately, the most abundant species of currant growing in the Pueblo Province, the wax currant, is not nearly as sweet or tasty as the common garden variety. As we might guess from the name accorded this species, *inebrians*, the berries have been used by certain western Indian tribes to prepare an intoxicating beverage. However, there is no history of this or of the making of any other alcoholic drinks among Puebloans. Archaeological evidence suggests that wild currants or gooseberries have been a dietary item of Puebloans for at least a thousand years and probably much longer.

Along with sumac, mountain-mahogany, and oak, the stout branches from currant bushes were one of the favored hardwoods for manufacturing both bows and arrows. In her book *Pueblo Crafts*, Ruth Underhill described the arrow-making process:

> Some of the Tiwa say they looked for stems which were not curved but had a jog in them for then the arrow flew straight but was hard to pull out. The maker scraped off the bark, then smoothed and evened the shafts by rubbing with sandstone. If the sticks needed straightening he steamed them over the fire till they were pliable, then passed them through a hole in a piece of horn with which he could pull them into shape. Then he rubbed them smooth on sandstone (Underhill 1979).

APACHE PLUME
Rose family
(*Fallugia paradoxa*)

As a medium-size, angularly branched shrub, Apache plume is fairly unremarkable until it blooms, which it does from spring through late summer. Then the plants become covered with white, five-petaled roselike blossoms, which are soon followed by an even more spectacular display of feathery pink seed plumes, each cluster resembling a miniature pom-pom.

Dense stands of these shrubs often grow where soil moisture runs deep, especially along the edges of arroyos, on adjacent sandy flats, and in disturbed sites. It's common along the Falls, Ruins, and Tsankawi trails at Bandelier and also at Jemez State Monument.

Apache plume is one of several shrubs that can produce slender straight branches and so was used by many Puebloans and their ancestors for making arrow shafts. The commonest use today is for yard brooms since the branches are sturdy and don't break easily.

FALLUGIA PARADOXA

Virtually all Rio Grande pueblos still associate Apache plume with being *the* plant for making rough outdoor brooms, yet at least one pueblo maintains a special indoor use. At Sandia Pueblo, if not elsewhere, many homes keep one or two Apache plume brooms behind their ovens because they believe the plant to exert a positive spiritual effect on the household.

The women of San Ildefonso used to steep the leaves in water and wash their hair in the mixture to promote its growth.
At Cochiti the petals may be eaten right off the plant to prevent stomach gas. "Just like Tums," as one Puebloan says.

The roots are long, wiry, and easily accessible in fresh erosion cuts on the sides of dry arroyos. As recently as 1970 Apache plume roots were being used at Cochiti for cord to tie fencing and make ramadas.

MOUNTAIN-MAHOGANY
Rose family
(*Cercocarpus montanus*)

Mountain-mahogany shrubs typically grow to six feet and sometimes much taller. The wedge-shaped leaves, an inch or so long, have notched margins along the apex but straight margins toward the base. Each small, dull, petal-less white flower of early summer produces a remarkable long, twisted plume that may hang on the plant until well into winter. When fully plumed, the plants often appear silvery from a distance.

These shrubs tend to grow in patches, especially on steep slopes, throughout the upper piñon-juniper and lower ponderosa pine ecozones. At Bandelier you will find them along the Frey, Upper Crossing, and Tsankawi trails.

The tough, thick branches of mountain-mahogany have been employed to manufacture a variety of implements—including fire-

CERCOCARPUS MONTANUS

stick spindles at Acoma and Laguna pueblos, boomerang-shaped sticks for killing rabbits at the Tewa-speaking pueblos, digging sticks at several other pueblos, and prayer sticks at Zuni—right up to the present. Tradition has it that arrows were once made from mountain-mahogany limbs, but this may have happened rarely or not at all. Sandia Puebloan Felipe Lauriano remembers discussing this years ago with his grandfather, who wondered how anyone ever found branches straight enough for arrow shafts since they almost never grew that way. However, the hard wood *was* carved for arrow points, as recorded at Acoma and San Juan pueblos.

Not much is known about medicinal applications of this plant. People from some of the northern pueblos once drank a mixture of ground leaves and cold water for a laxative, and the Acomas used to make some kind of a medicinal tea from the boiled leaves.

Perhaps the most widespread and well-known use of mountain-mahogany has been to achieve a reddish-brown color for dying moccasins and leggings. A number of pueblos have produced this dye made from boiled bark or sometimes the mashed roots. Charles Lange (1959) described how it was done at Cochiti: "Shreds of bark are placed in a half-gallon of hot water, to which a tablespoon of lime is added. The color is tested with a white corncob; the more lime used, the darker the color. The suds are spread over the buckskin until it reaches the desired shade."

THREELEAF SUMAC

Sumac family
(*Rhus trilobata*)

This rounded shrub grows up to six feet high, with the old growth branched and hard but the new growth straight and supple. Small yellow flowers appear in clusters along the branchlets in early spring, before the leaves come out. The leaves are in threes, and each leaflet is toothed and shiny green on top. The berrylike fruit is less than a half-inch in diameter, orange to red in color, and sticky, with short, glandular hairs.

This useful shrub is usually found on rocky slopes, as at Bandelier. But it obtains maximum size on the sandy bottomland of broad canyons at places such as Puye, Tsankawi, and other prehistoric cliff villages in the Pueblo Province.

Often called lemonade-bush, its delicious but acidic berries, which taste distinctly like lemons, are crushed and drunk as a summer refreshment at several pueblos today. Because of its affinity for

RHUS TRILOBATA

prehistoric fields, threeleaf sumac is thought to have been a managed plant—not domesticated but certainly encouraged and manipulated. One prehistoric Puebloan method of management may have been to set the bushes on fire. Fire stimulates shoot formation, as is the case for several other local shrubs, such as gooseberries and rabbitbrush. The new shoots are straight, long, and supple, but when dry they become hard and rigid, making them ideal for arrow shafts, among other implements.

Wicker baskets made from sumac are still crafted at Jemez Pueblo, but the shrub is no longer found much near the village. Patrick Toya of Jemez describes how some of the older folks still harvest berries and put them in tortillas to make a sandwich. He also says that in the old days the branches were shaped for infants' backboards and crib hoods, with cheesecloth netting then draped over them.

Sumac stem basketry fragments over two thousand years old were found during the excavation at Jemez Cave in the 1930s. One fragment was its natural color and another had been dyed black. In historic times sumac leaves were boiled with piñon gum, clay, and several other ingredients to provide a black dye effective on cotton or woolen cloth.

After seeing our slide show on wild plants at the Tesuque Pueblo senior center, some members offered that they used the crushed leaves of threeleaf sumac as a foot powder and a deodorant. The berries are also sometimes used for tea.

CANE CHOLLA
Cactus family
(*Opuntia imbricata*)

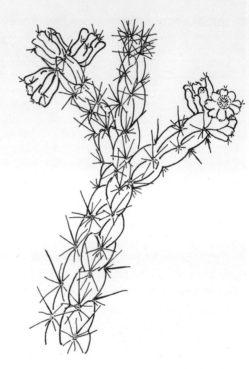

Cane cholla is a shrublike cactus with spiny, cylindrical joints or stems. The fleshy, grooved stems show the latticed pattern of the woody skeleton, which can be seen on dead portions of the plant. Various species of cholla in the Southwest may have yellow, orange, or purple flowers. In New Mexico the flowers of the most common species, *Opuntia imbricata*, are brilliant magenta, and the bright yellow fruit may persist on the plant until the year after the flowers have bloomed.

Cholla can't be missed. You can see it from almost every highway that passes through grasslands or woodlands in the Pueblo Province, from Taos to Isleta and west to Zuni. It also grows along nearly all the trails covered in this book, except those at Petroglyph National Monument.

West of the Rio Grande and not far from Petroglyph, at an

OPUNTIA IMBRICATA

excavated pithouse estimated to have been built between A.D. 1 and 500, no evidence of pottery or corn was found, so it is assumed that neither was known to the people living there. However, pollen analysis revealed that cholla was one of the most important wild plants utilized by the inhabitants at this site. Stone-grinding implements and large storage and cooking pits suggest that the people living here subsisted primarily on wild plants. Because of the lack of water, baking was probably the preferred method of cooking, followed by the hot rock method—a system of relaying heated rocks into a water-filled basket.

Historically, cholla is said to have been a famine food. The dethorned joints, flower buds, and fruit were all eaten, but only in time of great need. However, it may have been a common but forgotten staple. A 1915 record at Zuni relates that the fruit was eaten raw and stewed or dried for winter use. The dried fruit was ground into a flour, mixed with parched cornmeal, and made into a mush.

Nowadays, only the buds of this plant tend to be mentioned in cooking recipes. Like other green vegetables, cholla buds contain few calories. But a two-tablespoon serving contains as much calcium as a glass of milk. Not until the introduction of domestic cattle, sheep, and goats into the New World did milk become part of the diet of adults living here. In fact, American Indians are among those many cultures of the world whose bodies, beyond the period of infancy, fail to produce much lactase, the enzyme needed to properly digest milk. Fortunately for their family health, some determined Puebloan women continue to collect cholla flower buds with wooden sticks manipulated as tongs, rubbing the spines off with a rock, then boiling and putting them in cornmeal mush as fruit, with a little sugar. The woody skeleton has been used for walking sticks or to make small, light cages. The spiny stems are sometimes tied together to make impervious stockades, and on occasion the cuttings have been planted as living fences.

WOLFBERRY or TOMATILLO
Tomato family
(*Lycium pallidum*)

A spiny, scraggly branched shrub, wolfberry is rarely more than three feet tall in our area. The older branches are a dark reddish brown, the newer wood a glistening pale yellow. Mature leaves are leathery, pallid green, and appear in small clusters along the branches; the plant is deciduous in winter and during drought. The flowers, which blossom from May to June, are about an inch long, creamy green, and funnel shaped. Orange-red berries, about a half-inch in diameter, very much resemble miniature tomatoes. Ripening in July, the berries don't remain for long on the plant, as they are relished by birds and other wildlife.

In the province of the Puebloans, the preferred name for this plant is wild tomatillo, although that name may also be applied to a close relative of wolfberry, the groundcherry (*Physalis* spp.). Wolfberry is often associated with prehistoric ruins and can be seen on talus slopes below cliff dwellings at Bandelier and alongside the

LYCIUM PALLIDUM

pueblo ruins at Jemez State Monument. If you visit Petroglyph National Monument, look for it along the Cliff Base Trail.

It has been speculated that wolfberry is a camp follower, finding the disturbed soil of prehistoric ruins suitable for its growth (see chapter 5, "Indicator Plants as Living Artifacts"). Indeed, wolfberry seeds are a "normal" component of plant remains found in archaeological excavations at prehistoric sites.

In our study we found that nearly all contemporary Puebloans have at one time collected and used parts of this plant in various ways. From Matilda Coxe Stevenson's handwritten notes on the ethnobotany of San Ildefonso and Santa Clara (Stevenson 1912), we learned that the fresh or dried and reconstituted leaves of *Lycium* were applied to cuts and regarded as an excellent medicine. Of wolfberry's culinary use at Zuni Stevenson reported: "the berries are boiled, and, if not entirely ripe, they are sometimes sweetened.... The berries are also eaten raw when perfectly ripe." Volney Jones (1931) noted that the fresh berries were eaten at Isleta Pueblo in the 1920s. In 1978 a young doctor studying at Zuni Pueblo reported: "The root is soaked in water overnight. Bits of the root are then planted with corn to keep worms from eating the seeds and to make the corn plant grow fast."

At Acoma, Laguna, Isleta, San Juan, and Sandia, tomatillo berries may still be eaten fresh if ripe or sweetened and cooked if not quite ripe. But this shrub appears to be declining along the Rio Grande, and some Puebloans have told us that they no longer bother to collect tomatillos for the table.

BROOM SNAKEWEED
Sunflower family
(*Gutierrezia sarothrae*)

Low-growing, rounded broom snakeweed shrubs are bright green in late spring and early summer but turn into golden yellow globes when their tight clusters of tiny flowers burst into bloom later in the season. When crushed, the shiny, narrow gland-dotted leaves emit a turpentine-like fragrance.

This species is native throughout the arid West, but it thrives especially on moderately disturbed ground and gradually increases at the expense of grasses in fields that have a long history of heavy grazing, mainly because the resinous foliage is unpalatable to domestic livestock. Broom snakeweed is common along all the trails covered in this book except those above 7,500 feet elevation.

For Pueblo Indians throughout New Mexico this is one of the most important medicinal plants, presenting an enormous variety of applications. Hispanics, too, have discovered many medicinal uses

GUTIERREZIA SAROTHRAE

for snakeweed, some of them no doubt learned from the Puebloans shortly after the Spaniards first marched up the Rio Grande.

A tea made from the entire plant except for the roots is still used by Cochiti men for an emetic; for the same purpose Cochiti women prefer a weaker brew steeped from juniper twigs. People from Acoma, Laguna, Santo Domingo, and Santa Ana pueblos have used snakeweed tea as a cleansing emetic, historically sometimes to purify the body before a hunt. At Jemez, San Juan, and San Ildefonso the tea or its vapors has been used by birthing mothers or on the newborn following childbirth and at Zuni to relieve urinary retention or to increase physical strength. Rheumatism, rattlesnake bites, and eye problems have all have been treated with such a tea at various Rio Grande pueblos.

Sweat baths concocted from steeped broom snakeweed are associated with curing colds and sore throats at Santa Clara and fever at Isleta, and a liquid concentrate has been applied on sores, bruises, or aching muscles at Isleta, Acoma, and Zuni.

With curative powers so widespread among Indians and Hispanics it's surprising that to this day none of the active chemical compounds contained in snakeweed seem to have been commercially extracted from this plant for an ingredient in modern medicines. Plant researchers such as Dr. Stan Smith of Las Cruces are well aware that several of the compounds associated with snakeweed have potential pharmaceutical uses, yet we have found no reference to it in any recent medical literature. Snakeweed is considered to be a particularly undesirable weed on rangeland throughout the West, a shrub that ranchers cheerfully would jettison from their pastures. With all those resins, it would seem to be a prime candidate for intensive research by pharmacologists.

Clearly, snakeweed is a plant associated with good feelings at some pueblos. At Sandia each year people take bundles of wood mixed with snakeweed branches to the home of their *cacique*, the chief of the pueblo, during the twelve days following the winter solstice to serve as a blessing and for nurture.

RABBITBRUSH

Sunflower family
(*Chrysothamnus nauseosus*)

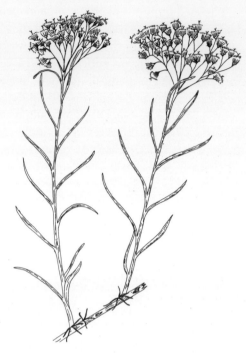

Golden borders along roads and arroyos throughout the Pueblo Province in fall owe their presence to rabbitbrush, or chamisa, as this shrub is called in most of New Mexico. Young twigs and leaves are covered with a fine white wool, giving the plant a grayish green look. Dense clusters of yellow-green flowers that turn to rich gold top this shrub, which can grow as tall as a person.

A deep root system enables rabbitbrush to tap underground moisture; thus, it grows along dry, sandy watercourses, often mixed with Apache plume, in the juniper-grassland and piñon-juniper ecozones. One of the earliest shrubs to green up in spring, rabbitbrush is heavily browsed by deer before most other plant foods are available. It's common along all the lower elevation trails at Bandelier as well as at Jemez State Monument.

The early Spanish recorded that Puebloans up and down the Rio Grande wore cotton garments decorated in various colors. We don't know exactly what colors, but if yellow was one of them, rabbitbrush blossoms surely were the source for the dye. Yellow dye

CHRYSOTHAMNUS NAUSEOSUS

for buckskin shirts and leggings as well as cotton and wool material would have been prepared by boiling crushed blooms, a practice that continued at many pueblos well into the twentieth century. At Zuni, yellow-dyed rabbitbrush stems were sometimes worked into baskets.

The record we have of rabbitbrush as food comes from San Felipe Pueblo, where the flower buds once were eaten. But medicinal uses have been described from most of the Rio Grande pueblos (though not from Zuni). A tea made by steeping the leaves seems to have worked for curing stomach disorders at several pueblos. This tea was also used as a gargle for colds at Jemez and as a bath for fever patients at Isleta; at Sandia ground dry leaves mixed with cornmeal were used in treating wounds.

Certain insects like to deposit their eggs on the twigs and stems of this plant, causing galls a half-inch in diameter to form. A liquid mixture of ground rabbitbrush galls was once employed at Cochiti to relieve toothaches and at Santa Clara for stomach problems.

The making of arrows or arrowheads from mature rabbitbrush limbs has been reported at Acoma, Laguna, and Isleta pueblos, but this certainly was not the preferred wood for arrow parts throughout the Pueblo Province.

Rabbitbrush, along with fourwing saltbush, has been named as a possible indicator plant of prehistoric fields. It does seem likely that this plant, with its deep roots, would be attracted to places where the remains of ancient water conservation devices still cause moisture to be retained at a depth, even though the cultivated plants are long gone.

SAND SAGEBRUSH
Sunflower family
(*Artemisia filifolia*)

FRINGED SAGEBRUSH
(*Artemisia frigida*)

BIG SAGEBRUSH
(*Artemisia tridentata*)

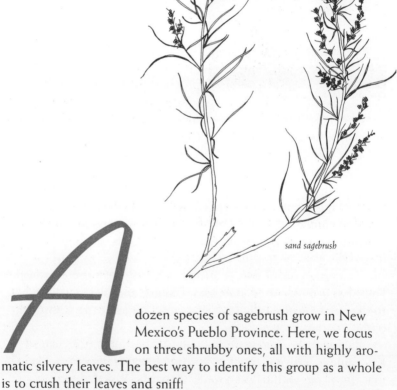

sand sagebrush

A dozen species of sagebrush grow in New Mexico's Pueblo Province. Here, we focus on three shrubby ones, all with highly aromatic silvery leaves. The best way to identify this group as a whole is to crush their leaves and sniff!

Sand sagebrush, which grows up to several feet tall, has an almost willowy form and threadlike leaves and is usually restricted to sandy places such as at Coronado State Monument or Petroglyph National Monument.

ARTEMISIA FILIFOLIA

The mat-forming fringed sagebrush seldom exceeds two feet in height and has tiny, crowded leaves. This species is the most widespread in our area; in fact, it ranges from Alaska nearly to Mexico in the West and also occurs in Siberia. Look for it along the lower-elevation trails at Bandelier and Petroglyph.

Big sagebrush is the tallest of the three, at a height of up to ten feet or more, and has distinctive three-toothed leaves up to an inch or so long. It's common in the northern Rio Grande Valley, especially around Taos, and can be found along the Falls and Tsankawi trails at Bandelier.

The principal uses of these plants are medicinal, largely owing to aromatic oils, including camphor, contained in sagebrush leaves.

All three species have been used by Puebloans as a cure for stomach disorders. A tea from the leaves was drunk at Picuris, most of the Tewa-speaking pueblos, and as far south as Isleta. The leaves could also be chewed directly, as at Cochiti and San Ildefonso, inducing vomiting and thus emptying the system of whatever might be causing the problem. Tewans used to steep a bundle of sand sagebrush plants in boiling water and apply it to the stomach as a hot compress.

With its camphor content, sagebrush has been highly valued in treating colds and coughs. Even today at Zuni big sagebrush is boiled in water and the steam inhaled as a decongestant; at San Juan Pueblo warm leaves may be applied to the neck to help a sore throat.

The leaves of big sagebrush are the most pungent of the three species and thus have been preferred for making medicine when a strong one is required. A recent article in the *Albuquerque Journal* showed an elder from Jemez Pueblo demonstrating how sagebrush may be used to fumigate and purify a home. Although sand and fringed sagebrush are plentiful around Jemez, big sagebrush doesn't grow there. Instead, the people collect it from Navajo lands many miles to the west. In fact, they call it "Navajo sage" and like to burn it and inhale the smoke for occasional depression. As one elder put it, "This plant is associated with Navajos who often live alone but are not lonely."

Use of sagebrush goes back a long way. A sprig of this plant was among the more-than-2,000-year-old remains excavated from Jemez Cave, and ingested sagebrush parts have been recovered at Anasazi sites in the Four Corners region. Much, much later, pioneers in Colorado made a tea from this plant for treating Rocky Mountain spotted fever.

GRASSES AND GRASSLIKE PLANTS

BROAD-LEAVED CATTAIL
Cattail family
(*Typha latifolia*)

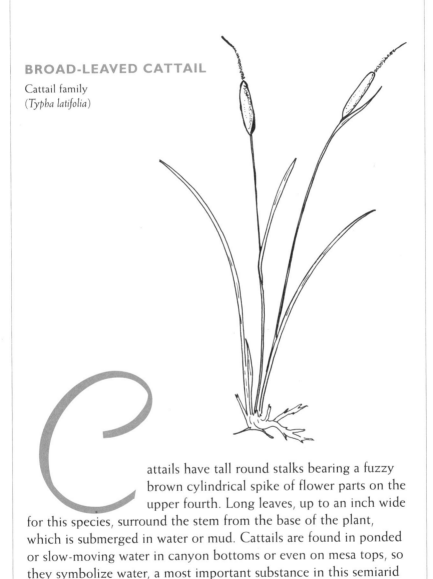

Cattails have tall round stalks bearing a fuzzy brown cylindrical spike of flower parts on the upper fourth. Long leaves, up to an inch wide for this species, surround the stem from the base of the plant, which is submerged in water or mud. Cattails are found in ponded or slow-moving water in canyon bottoms or even on mesa tops, so they symbolize water, a most important substance in this semiarid land. Thus, at most pueblos, various parts of the cattail are used in private and public ceremonies and dances.

TYPHA LATIFOLIA

All parts of the cattail—roots, shoots, pollen, seeds—are edible at some point in the growth cycle of the plant. Among the Acoma and Laguna the roots and tender shoots once were salted and eaten. At San Felipe the shoots were ground and mixed with cornmeal, particularly when food supplies were low. There is ample documentation to suggest that wetlands once were much more commonplace along the Rio Grande and that Pueblo Indians did not overlook this "supermarket of the swamp."

Cattails were extensively used as fiber and in house construction. During the excavation of Arroyo Hondo Pueblo a woven sleeping mat was found made entirely from a related species, narrow-leaved cattail (*Typha angustifolia*); another such mat was made of a combination of cattail and yucca fiber. At pueblos where cattail wetlands are still abundant, such as Isleta and Zuni, persons wishing to build in the traditional fashion find the stems of cattail useful in the construction of roofs: the stems are used as lathing over the vigas and thatch to support the mud or adobe cover.

INDIAN RICEGRASS
Grass family
(*Oryzopsis hymenoides*)

Indian ricegrass is a perennial bunchgrass that produces an open cluster of seed heads up to two feet tall in the spring or early summer; thus, it is known as a cool-season grass. The bell-like shape of the two bracts encasing each single seed helps make this one of the easiest grasses to identify, even in winter, after the seeds have dropped.

Ricegrass prefers dry, sandy soil. When protected from overgrazing, it becomes dominant on the windblown sands that stretch for great distances on either side of the Rio Grande. It extends up into the piñon-juniper ecozone, where it grows in pockets of sandy soil on the mesas and in stream valleys. Look for it along the trails at Jemez and Coronado state monuments and Petroglyph National Monument and on all the lower trails at Bandelier.

ORYZOPSIS HYMENOIDES

Ecologists believe that Indian ricegrass was once much more abundant on the open plains west of the Rio Grande but that severe overgrazing of the rangeland by sheep and cattle between 1870 and 1930 nearly eliminated it. In recent years, under more enlightened land management practices, this and other cool-season grasses have been returning.

Indian ricegrass was probably the most valuable wild cereal harvested in our area during prehistoric times, as evidenced by the remains of seeds unearthed from numerous archaeological sites.

Ricegrass seed continued to be regularly harvested and eaten well into the time when cultivated corn was the main subsistence food.

One reason for the value of ricegrass as a food would have been its availability for early harvest, usually by mid-June, well before most wild plants bear edible parts and months before corn was ready to pick. Equally valuable was the large size of its grain and the ease with which the seed and chaff could be kneaded and shaken from the stems. The grains were probably then tossed in a basket with hot coals and parched to remove the attached hairs. Ricegrass may have been the main source of plant food calories for Indians living on the sandy plains of New Mexico during the Archaic period. The archaeological record shows that by Anasazi times ricegrass had become a lesser component of the peoples' diet, although the richness of its protein and carbohydrate content certainly would have provided nutritional balance.

This grass is one of the few wild plants known to have been cultivated by Indians in recent times, though not in our area. Years ago the Paiute Indians living in Owens Valley, California, sowed and harvested large crops of it. Later in this century the Zuni ground the dark, plump seeds, combined them with water and cornmeal, shaped the mixture into balls, and steamed them for the table. Zuni Puebloans formerly sought out the grains of this plant in times of food shortage, but there is no record of ricegrass use by any of the modern Rio Grande pueblos.

COMMON REED
Grass family
(*Phragmites communis*)

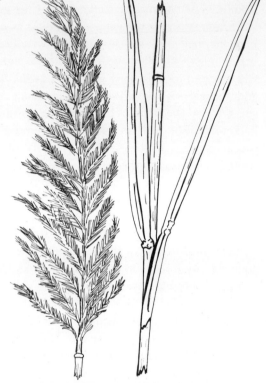

Few plants are more wide ranging than this species, which is native to every continent except Antarctica. Ten feet or more tall, its thick hollow stem—the culm—is topped by a graceful, silvery, foot-long plume of flower heads.

The common reed likes to have wet feet. It grows in dense thickets along streams, irrigation ditches, and marshes at lower elevations, so you won't see it from the trails covered in this book. Once it was plentiful along the Rio Grande clear up to the northern pueblos, but flood-control projects, irrigation, and other riverbank manipulation have virtually eliminated it from the river system north of the city of Albuquerque, except for a few patches growing on the banks of the Jemez River. It still thrives in the wetlands of Ojo Caliente west of Zuni Pueblo and along the Zuni River.

PHRAGMITES COMMUNIS

The record of human use of this grass is extensive. Charred reed culms dating to late Archaic times, before the birth of Christ, have been recovered from Jemez Cave; contemporary Jemez Puebloans recall that the reeds were still being used for roof insulation not too many years ago. And excavations at 600-year-old Arroyo Hondo Pueblo turned up a sort of corn crib composed of common reed mixed with the stems of willow and piñon.

The best-known historic (and, no doubt, prehistoric) use for reeds is for making arrow shafts, for which use this plant has been preferred over some other materials because of its lightness. To strengthen the arrowhead-tipped end and the lower notched ends of the shaft, short pieces of limbs from hardwood shrubs such as oak or mountain-mahogany usually were shaved down and inserted into the hollow reed. It's likely that people from all pueblos along the Rio Grande and its major tributaries, as well as the Zunis, crafted arrows from common reed at one time or another.

Minor uses of common reed by Puebloans during the past century include eating the roots, raw or cooked, perhaps in times of famine, and smoking the hollow reed filled with "Indian tobacco," an old practice at Zuni, among other pueblos.

LITTLE BLUESTEM

Grass family
(*Andropogon scoparius*)

SIDE-OATS GRAMA

(*Bouteloua curtipendula*)

DROPSEED

(*Sporobolus* spp.)

NEEDLEGRASS

(*Stipa* spp.)

Sand dropseed

Identification of grass species poses a special challenge to all but the most devoted fans of this group of plants. We won't go into the fine points except to note that all grasses have jointed stems called culms, which are usually round and hollow. The several grasses

ANDROPOGON SCOPARIUS

mentioned here are relatively tall—eighteen inches or higher—and are perennial bunch-type grasses. All these are common in the juniper-grassland and piñon-juniper ecozones throughout the Pueblo Province. Look for little bluestem especially on the Ruins and Tsankawi trails at Bandelier and for dropseed and needlegrass at Petroglyph.

Both little bluestem and side-oats grama have been associated with broom and brush making at several of the northern pueblos in the old days. The dried culms would be collected in late summer or early fall, then tied in bundles. The soft tip end of such a bundle would be used to sweep the floor of an adobe home. Another bundle might be tied in reverse and the short butt end used for a hairbrush or to brush spines off cactus fruit. The culms of dropseed grasses once were used at Zuni to manufacture mats for covering openings in adobe homes.

The greater cultural contribution of dropseed, however, was its grain, an important food in prehistoric times. Although the seeds are smaller and more easily processed than those of Indian ricegrass, they are still larger than those of most other grass species. So it's no great surprise that charred seed grains of both ricegrass and dropseed turn up regularly in early habitation sites, as well as in desiccated human feces from these sites. Thus, anthropologists are certain of the early dietary importance of these grasses. Needle-

BOUTELOUA CURTIPENDULA **STIPA NEOMEXICANA**

grass, including one elegant species called New Mexico feathergrass (*Stipa neomexicana*), also is large grained and has been associated with prehistoric habitations.

Because of their high cellulose content, grass leaves and stems are indigestible by humans, so only the seed grains have a dietary connection. The roots of some grasses are edible, but we know of no records of underground grass parts being eaten in our area.

HERBACEOUS PLANTS

NODDING ONION

Lily family
(*Allium cernuum*)

Nodding onion is the commonest of several wild onion species growing in the Rio Grande watershed and the most often sought for food. Each bulb typically produces four to six thin leaves in the center of which is a single flower stem a foot or so tall. The stem arches downward at the tip, from which droops a cluster of white to deep pink flowers.

This species doesn't grow much below the ponderosa pine ecozone, so you won't find it along the lower elevation trails covered

ALLIUM CERNUUM

here. Look for nodding onion in the woods along the Upper Crossing Trail at Bandelier.

Throughout North America people from all cultures have collected wild onion bulbs, and this species, ranging from coast to coast, is one of the favorites. High in vitamin C, the bulbs are eaten raw or used as a seasoning with cooked foods. Sometimes they are dried for winter use. A closely related group of plants, death camas (*Zygodenus* spp.), look a lot like wild onion and have a similar bulb, though they lack the characteristic onion smell. Death camas bulbs are extremely toxic to humans, and, reportedly, these plants have taken their toll among some Indian tribes. However, there are no records of Puebloans ever having made such a mistake, probably because wild plant species identification and naming was formerly so important to the educational process in Pueblo cultures.

Former medicinal uses of nodding onion bulbs appear to be minor. They include applying warm onions directly to the throat as a relief for sore throat at Isleta Pueblo and drinking onion juice for pneumonia at San Juan.

FALSE SOLOMON'S SEAL

Lily family
(*Smilacina* spp.)

Smilacina is commonly found with conifers, especially on north-facing slopes or in moist areas. The single stems rise from a perennial root stock and have the appearance of being jointed at the points where the oval, alternate leaves are attached. The leaves are somewhat thick and strongly veined; the berries are white with purple or green spots. Small white flowers are loosely clustered at the end of the stems. False Solomon's seal is common along streams at Bandelier and throughout the Jemez Mountains.

There appears to be only one published reference to its use in our area, a claim that the berries were eaten by the Tewa. Indigenous people outside the Pueblo Province have found the starchy root to be edible. Ethnobotanist Richard Ford has related that *Smilacina* leaves are collected, thoroughly crushed by hand, and

SMILACINA RACEMOSA

then cooked like a green vegetable at Picuris Pueblo. The crushing process is needed to break up tough cells or the cooked leaves will have the texture of thin leather. Picuris Pueblo is in remote, mountainous territory, where native plants tend to be collected prudently in season and eaten immediately or dried and stored for later use.

WILD BUCKWHEAT
Buckwheat family
(*Eriogonum* spp.)

Antelope-sage

Many species of wild buckwheat grow in our area. The most common of these, *Eriogonum jamesii*, is known as antelope-sage. This perennial has a basal whorl of grayish spoon-shaped leaves with stems up to a foot tall bearing ball-like clusters of creamy white flowers.

Antelope-sage favors dry, rocky slopes. At Bandelier look for the showy blooms in late summer and early fall at Tsankawi, along the Frey and Falls trails, and on the sides of Frijoles Canyon.

Several members of this genus are noted for containing saponin, a slightly toxic chemical that can lead to severe stomach irritation when ingested. Thus, a liquid concentrate made from wild buckwheat roots has been used by various Puebloans over the years as an emetic for the treatment of stomachaches. Other me-

ERIOGONUM JAMESII

dicinal uses from the past include the application of a powder derived from the stems on cuts to prevent infection at Picuris, a liquid earache remedy comprising powdered soaked roots at San Ildefonso, and an eyewash made from wild buckwheat roots at Zuni.

Wild buckwheat must have been a fairly important food plant in prehistoric times. Carbonized seeds and flower parts have turned up at many Anasazi hearth sites in eastern Arizona and southern Colorado and in desiccated human feces at Chaco and the Four Corners area.

DOCK
Buckwheat family
(*Rumex* spp.)

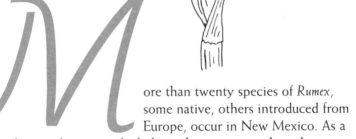

More than twenty species of *Rumex*, some native, others introduced from Europe, occur in New Mexico. As a group they are known as dock, but other names, such as sheep sorrel and wild rhubarb, apply to specific members of the genus. Most of the common docks are somewhat coarse, leafy perennials that flower early in the summer and exhibit dense, showy clusters of three-winged fruits.

Many of the docks found in the Pueblo Province occur in waste places such as abandoned fields or along roadsides. *Rumex hymenosepalus*, wild rhubarb or canaigre, is common in the sandy ground below the cliffs at Petroglyph National Monument and can be found at Coronado State Monument.

Both leaves and stems of dock have long been a food item for people from various Pueblo tribes, especially early in the season,

RUMEX HYMENOSEPALUS

when wild greens are at a premium. The leaves, brimming with vitamins A and C, may be treated like spinach, boiled and served as a table green with seasoning, or eaten raw in salads, although the high oxalic acid content of some docks makes them rather tart. This tartness probably is why a few Puebloans still like to toast the green leaves and save them for winter to mix with beans. Young stems are usually prepared much like rhubarb, and some pueblos used to favor a jam made by boiling down the stems. Other groups of Indians have ground up the seeds into meal for mush and breads, but this practice hasn't been reported from local pueblos.

Chemical analysis of at least one species of dock, the introduced *Rumex crispus*, has shown that the roots contain substances that inhibit bacterial growth. The Acoma, Laguna, and Zuni seem to have discovered this connection, for in times past they applied powdered roots to burns, sores, rashes, and skin infections. Patent medicines containing an extract of dock root once were sold as an antiseptic throughout the United States.

Wild rhubarb was among the several plants with high tannin content used at various pueblos in tanning rawhide. The Hopi have a tradition of using this species mixed with other plants to manufacture yellow dyes, but no dye use for dock has been recorded among the Puebloans of New Mexico.

GOOSEFOOT

Goosefoot family
(*Chenopodium* spp.)

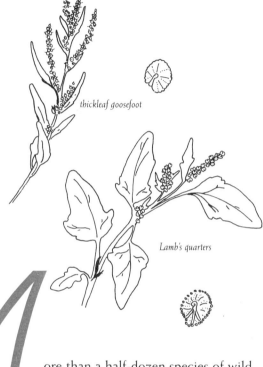

thickleaf goosefoot

Lamb's quarters

More than a half-dozen species of wild goosefoot and several more introduced ones grow in our area. Those with characteristic triangular or rhombic leaves are generally known as "lamb's quarters." All are weedy annuals with inconspicuous greenish flowers, and most of them are spindly plants that tend to have a mealy or glandular surface.

Like the seeds of amaranths, the next group of plants covered in this book, those of goosefoot germinate best after the summer rains in open, waste places. Usually by late July a few plants will be found around the ruins and along several trails in Frijoles Canyon at Bandelier.

Of the many annual herbs producing an edible and nutritious seed, the *Chenopodium* species are the most abundant and were cer-

CHENOPODIUM LEPTOPHYLLUM

tainly among the most important sources of food during early ancestral times, when large disturbed areas around the pueblos and the corn and squash fields would have provided fertile ground for thousands of these plants. Indeed, charred goosefoot seeds are associated with most Anasazi sites at which wild plant remains have been analyzed.

On an ounce-for-ounce basis, the seeds have roughly the same nutritive values in terms of crude protein and caloric content as Indian ricegrass, another major source of vegetable protein food in ancient times. Field experiments we undertook in the late summer proved that wild goosefoot seeds tend to remain on the plant in clusters and can quickly be stripped and harvested.

Goosefoot plants are similar to amaranth in many other ways, including the size, shape, and markings of their seeds and the microscopic appearance of their pollen. Prehistoric remains from the two groups of plants are difficult to distinguish, and archaeologists have often lumped the two together, calling them "cheno-ams." In any case, centuries ago, year in and year out, cheno-ams collectively provided a key wild plant source of calories and protein.

In more recent times goosefoot greens, not seeds, have become the part of the plant used for food. Residents of most, if not all, of the nineteen New Mexico pueblos at one time or another seem to have gathered and either boiled or fried the greens, but the practice has greatly declined in the past few decades.

AMARANTH

Amaranth family
(*Amaranthus* spp.)

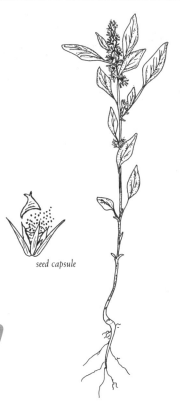

seed capsule

Wild amaranths found in our area are nondescript, coarse annual plants mostly under three feet tall. Typically, they bear clusters of inconspicuous bristly green flowers scattered along the stems where the leaves are attached and sometimes at the end of the stems. Several of the native species, sometimes called pigweed (among them *Amaranthus hybridus*, *A. retroflexus*, and *A. graecizans*) are associated with food use, but they are difficult to tell apart and are treated as a single plant group here.

Don't waste your time searching for these plants in a wilderness setting; they much prefer waste ground, cultivated land, or other disturbed places. By midsummer amaranths crop up around the Frijoles Ruins at Bandelier.

Economically, this genus of plants is divided into two major groups: grain amaranths and vegetable amaranths. The former, with

AMARANTHUS SP., TYUONYI RUIN, FRIJOLES CANYON.

immense and often colorful seed heads, is among the world's oldest and most important crops. For example, it has been recorded that, annually, some twenty thousand tons of amaranth grain were forced to be delivered in tribute to Montezuma, the Aztec emperor of Mexico in the 1500s. But the grain amaranths of tropical America don't seem to have made their way north to our area until historic times, and there is no evidence that they were ever cultivated or encouraged in gardens along the Rio Grande until quite recently.

Like their heavy seed head relatives, the so-called vegetable amaranths also produce tiny edible seeds but in far fewer numbers. Clearly, amaranth seeds have been an important source of food dating back at least to early Anasazi times. Seeds from these wild plants of New Mexico regularly turn up in the excavation of old ruin sites, often in food pits, sometimes mixed with other edible seeds. Amaranth seeds were even found in a bowl excavated from a pithouse on the lower Pajarito Plateau. That these seeds have been identified in desiccated human waste recovered from the Four Corners area and Chaco Canyon is proof positive of their role in prehistoric diets. They were still being collected and eaten or ground with meal and steamed at Acoma, Laguna, Zia, and Zuni pueblos well after the turn of the century.

Over time amaranth greens probably played an even greater dietary role—and with good reason, for the stems and leaves contain amino acids, vitamin A, caloric food energy, and various other essential nutrients comparable to and sometimes greatly exceeding those of domestic spinach. In fact, in a modern taste test conducted by scientists in Maryland, most participants claimed that cooked amaranth leaves tasted as good as or better than spinach, and we cheerfully attest to those findings. It's no wonder that until very recent times tender young amaranth greens have been universally used among the pueblos, sometimes dried for winter use but more often boiled lightly or boiled and fried with onions or other ingredients. In our area, amaranth stands among the "big four" of wild plant greens available for the picking throughout the summer, the others being beeplant, goosefoot, and purslane.

As mentioned previously, the old style of gardening encourages a mix of useful weeds among the standard crops; thus, it is mainly those who still garden this way who regularly harvest amaranth greens. Among them are Puebloans such as David Chavez at Zuni, who collects the greens growing among his corn and squashes "for spinach" when the plants are young, or Ramos Oyenque at San Juan, who cuts his field amaranths before they flower and may squeeze them dry into "biscuits," then store them until winter, when he and his family will eat the hydrated cakes with onion or garlic.

Finally, one species of amaranth, *A. cruentus*, probably introduced by the Spanish into this area in historic times, has been used in the Southwest as a dye plant. At Zuni as recently as forty years ago this tall, red-flowered amaranth was encouraged to reseed and spring up in some of the old waffle gardens in town or in the irrigated fields along the river. Then, the blossoms of the plant would be ground to a fine meal and used to color their *piki*, or wafer bread, red. Perhaps this is still done today at Zuni. The same practice was formerly attributed to the people of San Ildefonso Pueblo, but there is no evidence of dye use at any pueblo located along the Rio Grande in modern times.

FOUR-O'CLOCK

Four-o'clock family
(*Mirabilis multiflora*)

This handsome sprawling or bushy plant displays masses of inch-long magenta-purple flowers protruding from papery floral cups. Leaves are dark green, thick, and deltoid in shape. Four-o'clock is often found along game trails and under piñon and juniper trees and is conspicuous along the Ruins and Tsankawi trails at Bandelier.

Evidence of prehistoric use of *Mirabilis* has been found at the ancient Fresnel Shelter site in New Mexico. As to more recent use, the very large roots and other parts of the plant are reported to have been ground and used medicinally by nearly all the pueblos, although the illness for which they were used and the curative techniques may have varied from village to village. At Santa Clara a small portion of the ground root was mixed with water and taken for colic or relief from distended stomachs. It was also a remedy for

MIRABILIS MULTIFLORA

eye infection and a rubbing compound to alleviate sore muscles. At Zuni the powdered root was once mixed with flour to make a bread to decrease the appetite. The Acoma and Laguna dried the leaves for use as smoking material. These uses indicate that the plant has a sedative property, but to our knowledge the effective ingredient has not been isolated or identified.

Some researchers have suggested that four-o'clock is an indicator plant for ruins and farm gardens; however, that proposal does not seem to hold for sites in the Rio Grande Valley or Pajarito Plateau.

COMMON PURSLANE

Purslane family
(*Portulaca oleracea*)

With its flat, fleshy leaves and nearly prostrate, spreading growth form, this small annual herb is easily recognized. In midsummer common purslane has tiny yellow flowers in its leaf axils that then produce capsules filled with shiny black seeds. Common purslane originated in Europe and Asia, and has spread into temperate and tropical areas around the world; varieties of it have even been cultivated in some countries overseas. It grows best in open, disturbed ground and can be found along the Frijoles Ruins Trail at Bandelier.

Common purslane probably arrived in the Southwest during historic times and has subsequently nearly replaced the native purslanes. Purslane seeds recovered at archaeological sites and in coprolites associated with these sites come, for the most part, from

PORTULACA OLERACEA

the very similar indigenous notchleaf purslane (*P. retusa*). In any case, all evidence points to purslane being one of the most important wild food plants of Puebloan ancestors for at least a thousand years.

Zuni Pueblo seems to be the only one with a history of recent seed use. Frank Cushing told of how the Zunis prepared the seeds in the early 1900s: "This plant bore plentifully a small, black, very starchy and white-kerneled seed. It was gathered by pulling the plants just before the seed had ripened, then drying and threshing them either by agitation or by pounding them over mats or screens" (Cushing 1920). Presumably, the seeds were then mixed with those from plants such as sunflowers or Indian ricegrass for a mush or ground into flour for breads.

Though wild seed harvest may be out of vogue at most pueblos, the collection and preparation of purslane greens certainly are not. The succulent stems and leaves, high in vitamins A and C, may be boiled (then sometimes fried) and eaten alone. They're also mixed with peas or beans or added to a stew. Common purslane can grow so profusely in and around gardens that many Puebloans still like to slowly dry some of their wild crop and save it for winter meals.

Various Indian tribes, among them the Navajo, Acoma, and Laguna, have a medicinal tradition for purslane, the latter Puebloans at one time using a tea from the vegetative parts to cure diarrhea and as an antiseptic.

WESTERN TANSY MUSTARD

Mustard family
(*Descurainia pinnata*)

Tansy mustard is one of the earliest plants to appear in spring. A weedy annual, it grows from one to three feet high, branching more as it gets older, depending on the growing conditions. Leaves of this plant are finely dissected, filigree-like, with the dense, fine hair on younger plants giving a gray color to the foliage. Small, yellow, four-petaled flowers bloom profusely along the outside stems so that a field of tansy is a field of yellow. The seed pods are narrow, about a half-inch long, on thin half-inch-long spreading or upturned stalks.

Western tansy mustard is a common native weed of the Southwest. Its seeds and greens have frequently been identified as food items, both from prehistoric times and the present. Some local species of *Descurainia* were introduced from Europe; other varieties may be crosses between new- and old-world species. All are

DESCURAINIA PINNATA

weeds in the sense that they grow on roadsides, in old fields and gardens, and on disturbed or overgrazed lands.

The western tansy thrives at Coronado State Monument, mostly within the ruins of ancient Kuaua Pueblo. It and other mustards can be seen on talus slopes beneath cliff dwellings at Frijoles Canyon and Tsankawi. When growing on bottomlands, western tansy may be an indicator of prehistoric fields, but the introduced mustards are more likely to indicate historic land use, particularly overgrazing. Indeed, *Descurainia* is considered an indicator plant because it frequently occurs in the heavy soils caused by the weathering of adobe mortar or plaster.

Tansy mustard is used to produce pottery paint and is prepared in much the same way as *guaco* (the paint made from Rocky Mountain beeplant). Descendants of the former inhabitants of Pecos Pueblo have been successful in duplicating the lead glaze designs on pottery once used there: seeds of tansy mustard are ground in a mortar until an oily liquid forms; this liquid is then mixed with iron pigment and used to paint patterns on the pottery.

ROCKY MOUNTAIN BEEPLANT
Caper family
(*Cleome serrulata*)

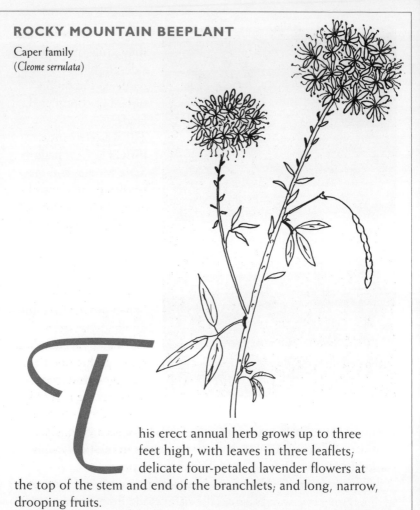

This erect annual herb grows up to three feet high, with leaves in three leaflets; delicate four-petaled lavender flowers at the top of the stem and end of the branchlets; and long, narrow, drooping fruits.

In some years beeplant can be found in early summer along the Ruins Trail at Bandelier National Monument, through the old fields leading to the village of Tyuonyi. Beeplant seeds germinate and produce new growth year after year in prehistoric canyon bottom fields such as the ones below Tsankawi Mesa and in old fallow fields of modern pueblos. Indeed, the presence of these plants is often an indicator of ruins and ancient farming plots.

A welcome volunteer in the fields and gardens of Puebloans in historic times, beeplant is collected when young and only a few inches high. When eaten as a green vegetable, it tastes alkaline and

CLEOME SERRULATA

bitter unless the cooking water is changed several times. An alternative method employed by the people at San Juan Pueblo to remove the bitter flavor has been to boil the greens with a piece of corncob. Supplementing meals with beeplant greens continues to be so widespread that many Puebloans refer to it as "wild spinach" or "Indian spinach."

As the plants grow older, the leaves may be collected and cooked down into a thick, dark paste, then sun dried on a board or stone to form small cakes. The cakes can later be broken and eaten with cornmeal mush or fried with fat. After the plant further matures, the seeds may be gathered, dried, and stored for winter use. Also, the whole plant—flowers, seed, leaves, and stems—can be ground, mixed with cornmeal, and baked in ashes into "cleome cornbread."

Warm cooked beeplant leaves were once used by the San Ildefonso people as poultices for sores and wounds. A tea brewed from the leaves was said to relieve stomach disorders.

The most distinctive and important use of this plant was as pigment: the reconstituted cakes of dried beeplant supplied the black designs on the white-slipped pottery so characteristic of this region. It also was the source for black designs on baskets crafted by various pueblos, as at Santa Clara, where most of the tray-type baskets were traditionally ornamented with black-and-red geometrics. Beeplant paint has endured on pottery, baskets, and other artifacts since at least the fourteenth century. This tradition has been maintained by Puebloan artisans up and down the Rio Grande; they usually refer to the beeplant pigment as *guaco*.

Seeds and pollen of beeplant have been found in excavations from prehistoric sites at Bandelier National Monument and in the Rio Grande Valley. We know for certain that its seeds were eaten during prehistoric times, for they have turned up in coprolites at several Anasazi ruins. In fact, one researcher has stated, "The ubiquity of beeweed pollen in paleofeces may indicate that this plant was consumed very commonly and may have been manipulated highly by humans" (Minnis 1989).

This plant has been of such economic importance that it is still named along with the main cultivated plants in songs of modern Puebloans living along the Rio Grande.

DOVEWEED
Spurge family
(*Croton texensis*)

This one-to-two-foot annual has narrow, drooping, lance-shaped leaves covered with fine white hairs, giving the plant a silvery look. Doveweed is *dioecious*, that is, the male and female flowers are on separate plants. Small clusters of flowers lacking petals appear in late summer. Each female flower produces a round fruit capsule containing three large seeds. Mourning doves flock in to devour the fruits in fall, providing this plant its common name.

Doveweed, mainly found in the piñon-juniper ecozone, grows best in disturbed soils and is considered an indicator plant for archaeological sites. It thrives among the ruins in Frijoles Canyon and can also be seen growing along the trails at Petroglyph National Monument.

Doveweed contains croton oil, a known cathartic for humans. This quality, coupled with the likely abundance of the plant in dis-

CROTON TEXENSIS

turbed places around ancient pueblos and cultivated fields, must have led to the discovery by ancestral Puebloans of some useful medicinal applications. We don't know how far back in time the plant was used, but in the past few decades doveweed tea has been drunk by Zunis for stomachache, by the Jemez for body and headaches, by the Isleta for laxative treatments, and by the Acoma, Laguna, and Zuni for general cathartic cleansing. The people of Acoma also have ground the ripe seeds to make a powder for skin sores. At Acoma and Isleta ear trouble has been treated using seeds or leaves placed in the ear. San Juan and other Tewa-speaking Puebloans may still boil the plant and bathe in the liquid to relieve rheumatism or for general well-being.

Other cultures have found that this plant, with its toxic croton oil, can be used to repel insects from the house or garden. Doveweed just doesn't smell very good!

GLOBE-MALLOW
Mallow family
(*Sphaeralcea* spp.)

Globe-mallow is common on roadsides and well-traveled paths of the Pajarito Plateau and Rio Grande Valley. This perennial seems to prefer disturbed land and, once established, endures for many years. Look for this plant along the Ruins Trail at Bandelier, at Jemez State Monument, and at Petroglyph National Monument.

The fruits of all globe-mallows are edible. Their pollen and seeds have been found at Chaco Canyon, Pecos, and on the Rio Grande near Zia and Santa Ana in archaeological contexts, suggesting that they were regularly eaten. In fact, globe-mallow seems to be one of the most widely used plants during this Anasazi era.

Utilization of this plant definitely has carried over into historical times. The people of Santo Domingo have boiled globe-mallow and added it to gypsum as a glue for calcimine house paint. At Taos

SPHAERALCEA COCCINEA

the pulp of the plant has been mixed with mud to make very hard floors; the root was also pounded and mixed with a little saltwater as a poultice to draw infection from sores and boils. A similar use is known for other pueblos: at Santa Clara, globe-mallow may be rubbed on sore muscles (because of the fine, stiff, star-shaped hairs on the foliage, the plant probably irritates the skin and brings blood to the affected area). At Picuris the roots of scarlet globe-mallow (*S. coccinea*) once were pounded into a pulp, mixed with water, and plastered over broken or fractured bones, solidifying into a hard cast.

Other medicinal properties have been attributed to globe-mallow, but medicine for sore eyes, as in one of its colloquial names, "sore-eye poppy," is definitely *not* one of them. This name probably is a result of children having eaten the buds or fruits of the plant and then irritating their eyes by rubbing them with hands covered with the fine but scratchy star-shaped leaf hairs.

PRICKLY PEAR
Cactus family
(*Opuntia phaeacantha* and similar species)

HEDGEHOG CACTUS
(*Echinocereus* spp.)

PINCUSHION CACTUS
(*Coryphantha* spp.; *Mammillaria* spp.)

In addition to the tall cholla (covered under the section on shrubs), many species of lower-growing cacti occur throughout the Rio Grande watershed. Although individual species are difficult to distinguish, most of the herbaceous cacti in our area fall into one of three main groupings. The prickly pear group has flat, stout-spined stems typically arranged in a jointed series of pads. The stems of hedgehog cacti are cylindrical and conspicuously ribbed, and they

OPUNTIA PHAEACANTHA

usually grow in clusters. The pincushions, the smallest of the three, sometimes consist of just a ball or single ribless cylinder protruding above the ground but may also be found growing in clusters. Flowers among these three groups of cacti come in many shades of yellow, pink, or red.

The various parts of these cacti have principally been used for food, but at Zuni the prickly pear fruit may be dried, ground, and dissolved in water with a chunk of dried beeplant to make dye. The mixture creates a reddish shoe polishlike substance that has been used on moccasins. This same prickly pear beeplant combination has also been used to dye thread or weaving fiber.

All prickly pear cacti in our area produce edible fruits. The fruits of the smaller, ground-hugging species are dry and plain, but the tall, larger-padded species yield succulent, fig-size red fruit called *tunas*. Through a simple, timed experiment done on a species with larger tunas growing downhill from a prehistoric pithouse village it was discovered that one person could collect nineteen pounds of fruit with a gloved hand in twenty minutes; a buckskin-mitted hand would have served the ancients just as well. Prickly pear tunas were one of the few sweets to which southwestern American Indians had access before the arrival of Europeans.

Over the ages the tunas of prickly pear have been collected and eaten at all of the pueblos. At some pueblos tunas are still collected, rubbed in grass or singed over a fire to remove the spines, and eaten raw as a snack.

If the spines are removed, the pads of prickly pear are also edible, boiled or raw. Because the flesh, like that of okra, is mucilaginous, a dish prepared from cactus pads was probably combined with dry ingredients such as cornmeal. Cultivated spineless prickly pear pads are now sold in neighborhood markets in the Southwest and are frequently an ingredient in modern Santa Fe cuisine.

Pincushion cactus stems used to be eaten by Puebloans as a snack. The spines were burned off and the cactus eaten whole or cut in half and the flesh scooped out and eaten. Unfortunately, many species of pincushion cacti are becoming endangered because of illegal gathering by irresponsible collectors.

Hedgehog cacti, with their larger and probably tougher stems, were sometimes roasted in a pit. Years ago ethnobotanist Volney H. Jones recorded the eating and preserving of the fruits of one hedgehog species, the spectacular *Echinocereus triglochidiatus*, known as claret cup cactus, at Isleta Pueblo: "The pulp is prepared in a variety of ways. It may be sliced and baked as squash is prepared. A sweet pickle is made by baking it with sugar. Cakes and candy are made from it in much the same way" (Jones 1931).

WILD PARSLEY
Carrot family
(*Cymopterus* spp.)

WILD CELERY
(*Cymopterus fendleri*)

WAFER PARSNIP
(*Cymopterus bulbosus; C. montanus; C. purpureus*)

MOUNTAIN PARSLEY
(*Pseudocymopterus montanus*)

wild celery

A bright green spot in the dried grasses of New Mexico in early spring, wild celery is a diminutive herb about one or two inches high and is easily missed until you learn to look for it. The minute yellow flowers are in *umbels*, that is, gathered at the ends of short

CYMOPTERUS FENDLERI

stalks. Below the flowers are the glossy, deep green dissected leaves. It is said that when the plant flowers, the air around it has a fresh scent. Certainly, the leaves have the fragrance and distinct taste of celery. Even the crunchy root tastes a bit like celery.

Other species that often grow in the same habitat are generally referred to as wafer parsnip. They have a similar growth form but the leaves are dull, the flowers are usually purple, and there is little or no smell of celery. Wafer parsnip is probably named for the damp cracker texture of its root. Collectively, these plants are often referred to as wild parsley.

Perennial, they are much sought after in March, April, and May, months when stored vegetables are running out and domesticated plants have not yet been planted or sprouted. These plants are generally found among pebbles and cobbles on the sandy hills and mesas of the juniper-grassland and lower piñon-juniper ecozones. Wild celery blooms in April among the volcanic rocks along the trail to the mesa at Petroglyph National Monument, and both wild celery and wafer parsnip are common on the stony hills adjacent to the highway from Santa Fe to Albuquerque. After the winged

PSEUDOCYMOPTERUS MONTANUS

seeds set in late spring or early summer, the leaves become dry, brown, and curled, nearly disappearing until the following spring.

People from all the pueblos and others who have learned from Puebloans regularly collect the leaves of wild celery in the spring. These leaves are delicious simply chewed as a mouth freshener or appetizer. Most often they are used fresh or dried as seasoning in beans, stews, and salsa or sprinkled on meat. At Cochiti years ago we helped collect the leaves, which were rinsed in saltwater and served as the perfect condiment to a red chile dish. One lady at the Tesuque Pueblo senior center told us she felt that wild celery was helpful for her arthritis.

At San Juan the peeled roots of wild celery and wafer parsnip are also relished in the spring. Digging up these roots is no easy task, as they seem to seek the underside of larger cobbles, probably for the moisture contained there. As the plant is dug up, the first storage root encountered is about six inches below the surface. It can be from four to eight inches long and has a thin, stringy end that connects to another, deeper storage root. Needless to say to anyone familiar with New Mexico soils, the deeper the hole, the harder the digging. The second root is probably rarely excavated but rather left for future production. Most often, just the leaves of this plant are picked, leaving the flowers, seeds, and roots to propagate anew.

A close relative of these plants, mountain parsley, usually found growing at higher elevations in the ponderosa pine or mixed conifer ecozones, has parsleylike leaves and similar-looking flower clusters that may be either yellow or purple—sometimes both on the same plant! Like the others, leaves from mountain parsley are known to have been eaten fresh or used for flavoring in salads or soup.

MILKWEED
Milkweed family
(*Asclepias* spp.)

Nearly a dozen species of milkweed occur in the New Mexico Pueblo Province. All but one have milky sap and can be identified by their flowers, composed of five erect, hoodlike structures and borne in showy clusters. Also distinctive are their large fruit pods, which enclose seeds with long silky hairs. Leaves of the different species have a variety of shapes, and the flowers range from greenish white to pink or orange; we won't try to sort the species out here. Milkweeds can be seen in the meadows of the Upper Crossing Trail at Bandelier and along the roads to Jemez State Monument and Taos and Picuris pueblos.

In times past young, green milkweed pods have been sought out by almost all Pueblo people, especially youngsters, to be eaten raw. Tewa-speaking people also ate the uncooked roots, probably from one of the species that forms fleshy tubers underground. At

ASCLEPIAS ASPERULA

Picuris Pueblo young broad-leaved milkweed (*Asclepias latifolia*) plants are said to have been used for flavoring meats. And at several pueblos the dense, milky juice that exudes from cut stems was allowed to thicken and then used for chewing gum or mixed with cottonwood fluff and chewed, as at Sandia Pueblo. The narrow-leaved species, *Asclepias subverticillata*, would have been avoided as a food item since it is poisonous.

The genus *Asclepias* was named after an early Greek authority on medicinal plant properties. More recent laboratory analysis has demonstrated that the sap from milkweeds contains various chemicals that can act as a stimulant or a diuretic or have other medically beneficial attributes. Indeed, this broad spectrum of therapeutic effects has resulted in the use of milkweed by many cultures for hundreds of years. Medicinal uses attributed to Puebloans in the past are legion. A tea made from plant parts was a remedy for stomach troubles at Acoma and Laguna and as a treatment for fevers and coronary problems at San Juan. Other Tewa Indians also prepared a tea of ground milkweed roots to help chest pain victims. At Isleta the people used to inhale powdered leaves and stems to relieve a stuffy nose. Today, most Puebloans are more likely to purchase dried milkweed parts at local markets with health food sections than to seek out plants in the wild.

Fibers, called *bast*, from the stems of milkweed have been identified in prehistoric textiles associated with our area. One of the earliest ethnobotanical researchers in New Mexico found that Tewa-speaking people of the Rio Grande were still making string and rope from these fibers. However, at Zuni it was the silky seed fibers that once were spun on a hand-held wooden spindle into yarn and woven into fabric, especially for use by dancers.

BLUE TRUMPETS
Phlox family
(*Ipomopsis longiflora*)

Named for the shape and color of its showy pale (sometimes white) flower, this annual frequents sandy soils throughout the piñon-juniper ecozone. Few other plants in our area exhibit a delicate flower with such a long tube—up to two inches in length. Toward the end of summer look for blue trumpets blooming along the Ruins Trail and the lower part of the trail to Tsankawi at Bandelier.

The known uses for blue trumpets are strictly medicinal. Various Tewa Puebloans of the north used to make a lather from dried flowers and leaves and apply it to body sores or to the face to cure headaches. Among the Tewa-speaking Indians, at San Juan the curative uses of blue trumpet extended to fevers, swellings, and even broken limbs.

IPOMOPSIS LONGIFLORA

To the south the Acomas are reported to have drunk a tea made from the roots when an emetic was desired. Until recently, Zuni Puebloans have used the lather from dried flowers to remove hair from newborn babies. The Zuni also appear to have once used lather from a related plant, the many-flowered ipomopsis (*Ipomopsis multiflora*), as a remedy for headaches.

SCORPIONWEED

Waterleaf family
(*Phacelia* spp.)

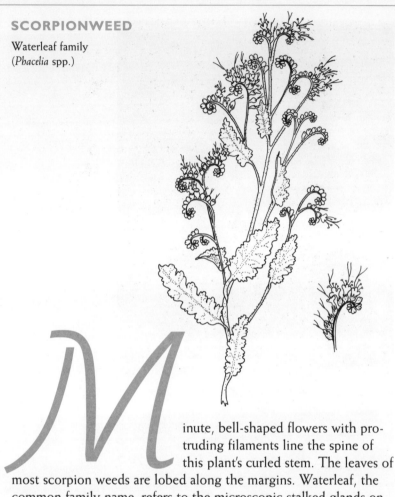

Minute, bell-shaped flowers with protruding filaments line the spine of this plant's curled stem. The leaves of most scorpion weeds are lobed along the margins. Waterleaf, the common family name, refers to the microscopic stalked glands on flowers and foliage, which resemble droplets of water. Seeing these glands requires a hand lens, but their presence gives the plant a sticky feeling and a glistening appearance.

One species of scorpionweed, *Phacelia corrugata*, thrives in the dry barren soils alongside the ruins of Kuaua Pueblo at Coronado State Monument and around the cave sites at Tsankawi at Bandelier, as well as at other archaeological sites. It appears to be an indicator plant for these specialized habitats on the Pajarito Plateau. An association with archaeological sites in Chaco Canyon also has been

PHACELIA CORRUGATA, KUAUA KIVA, CORONADO STATE MONUMENT.

reported. Elsewhere in the Southwest scorpionweed is considered an indicator of prehistoric fields, but that does not seem to be the case in our area.

Various species of *Phacelia* were used medicinally at Acoma, Laguna, San Ildefonso, Santa Clara, and, most recently, at Zuni. The powdered root or leaves are mixed with water and rubbed on sprains, swellings, and rashes. This is an interesting group of medicinal plants with unique habitat requirements, and they deserve more study.

BEE-BALM

Mint family
(*Monarda menthaefolia*)

The highly aromatic flowers of bee-balm, also known as mountain oregano, are magenta or purple and occur in terminal heads above the two-to-three-foot stems. The foliage is bright green, and the stems are square, like those of most other mints. Bee-balm is usually found in open ponderosa forests and is quite common on the Upper Crossing Trail and occasionally along the Ruins Trail at Bandelier.

Medicinal uses are recorded from nearly all the pueblos. At Santa Clara the dried leaves were crushed, mixed with water, and drunk to alleviate stomachache. At San Juan a decoction of the root was administered to heart attack victims. Other Puebloans have used this plant in solutions for cleaning open sores and wounds, for an eyewash, to cure headaches, or to break a fever.

MONARDA MENTHAEFOLIA

Like a lot of other folks, people from most of the pueblos still use aromatic bee-balm leaves, fresh or dried, sprinkled on meats or as a seasoning for soups and stews. At Zuni wild plants are sometimes transplanted into home gardens to be grown and used as an herb for flavoring various dishes.

JIMSONWEED
Tomato family
(*Datura meteloides*)

Jimsonweed can be recognized by its sprawling growth form and large leaves, and even more surely by the six-to-eight-inch trumpet-shaped white flowers. As one describer put it, jimsonweed, which usually grows in desertlike places, looks like a "fugitive from a tropical rain forest."

This annual plant is not well established in the Southwest. It favors habitats almost exclusively created by human or natural disturbance, such as roadsides, vacant city lots, or eroded arroyos at lower elevations. At Bandelier you find it growing alongside the Ruins Trail near Long House and Talus House. It is also one of the distinctive plants of Puye Cliff Dwellings, where it's been around for a long time. In fact, a photo of Puye published in *National Geographic* just after the turn of the century clearly reveals a number of these plants.

DATURA METELOIDES

But an association with archaeological sites in southern and western New Mexico extending north to the Pajarito Plateau is much older than that, indicating that jimsonweed must have been used extensively, probably for medicinal purposes, by Puebloans for centuries. Thus, it's surprising that, except for a single obscure reference to Tewan medicine, there is no record of any use by Rio Grande Puebloans within historic times. Perhaps the people here long ago learned from experience of the dangers this species holds. All parts of this plant contain highly concentrated poisons—even to the touch. Certainly it's a good plant to avoid.

The alkaloids in jimsonweed can produce a numbing effect or even death. West of the Rio Grande some of the more isolated villages have a history of grinding the roots, leaves, or flowers and applying the powder to wounds or as a poultice for skin infections, probably as an anesthetic or analgesic—under the direction of local experts. Jimsonweed has important ceremonial associations for various Indian tribes living in the western states, hence, the other common name for it is sacred datura.

GROUNDCHERRY
Tomato family
(*Physalis* spp.)

Groundcherries are just one of many members of the tomato family that has great economic importance. Others include chile and bell peppers, Irish potatoes, eggplants, tobacco, and petunias, as well as many medicinal plants.

Several species of groundcherry grow in the Pueblo Province. The three principal ones are usually well under three feet tall and have nodding, five-angled yellow flowers with dark centers. After flowering, they are even more easily recognized by their papery inflated husks, which resemble miniature lanterns, each enclosing a single pea-size berry.

Groundcherries occur in canyon bottoms, slopes, and mesas, generally at lower elevations and often in somewhat disturbed soil. Look for them around the ruins in Frijoles Canyon at Bandelier and

PHYSALIS HEDERAEFOLIA

at Jemez State Monument. Archaeological evidence throughout the area makes clear that groundcherry fruits have been collected and eaten for at least 1,100 years, probably much longer. The plants seem to have a strong link with long-abandoned fields, and some anthropologists suspect that groundcherries were actually cultivated or at least encouraged among other crops during Anasazi times.

All modern Puebloans who have an interest in wild plants know this one and its edible berry. When ripe, the orange or red berry may be eaten raw or cooked and used as a flavoring. If not grown at home, the fruits, commercially grown husk tomatoes, may sometimes be purchased at vegetable markets. Recently fruits from wild groundcherries were selling for $3.09 a pound at a trading post near Zuni Pueblo.

The Zunis, who seem to have a knack for making condiments from wild plants, have a special attachment for groundcherries, or *tomatillos*, as they call them, as well as for other plants that produce tomato-like fruits. Years ago it was reported that these plants were being cultivated by Zuni women, who boiled the ripe berries and ground them in a stone mortar with raw onions, chile, and coriander seeds for a kind of salsa. Over the past several years Zuni High School students in Patricia Allen's language arts and composition class have developed a series of essays on uses of native materials at Zuni Pueblo. One essay describes the making of groundcherry salsa: "Peel off the husks and put the cherries in a pan and roast them in an oven for about 20 minutes. Then mix chile, coriander and onions in the blender to season the salsa. Fresh onions are often dipped into the salsa and eaten like chips and dip. This recipe is famous in the Zuni area (Allen 1989)."

HORSE-NETTLE

Tomato family
(*Solanum elaeagnifolium*)

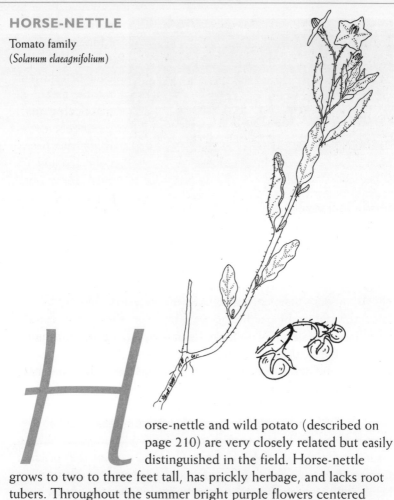

Horse-nettle and wild potato (described on page 210) are very closely related but easily distinguished in the field. Horse-nettle grows to two to three feet tall, has prickly herbage, and lacks root tubers. Throughout the summer bright purple flowers centered with five yellow-orange stamens accent the dull green leaves. Later, a juicy, globelike berry is produced—green or yellow at first, turning to orange, then brown or black in the fall.

This is another plant that prefers waste ground along roadsides and other recently disturbed sites. It thrives among the ruins at Coronado and Jemez state monuments and can be seen at Petroglyph and along the Ruins Trail at Bandelier.

The roots of horse-nettle possess chemicals that seem to have both antiseptic properties and the ability to draw fluid from tissue. The Zunis must have discovered this long ago, for they once used

SOLANUM ELAEAGNIFOLIUM

the roots on snakebites and for toothaches.

Rio Grande Pueblo Indians have drunk a brew of plant parts for stomach sickness and nursing mothers to sustain milk flow. At Isleta the berries, boiled into a syrup, were thought to have a laxative effect. The men of Cochiti are said to have dried the berries and used the powder as a kind of snuff, like tobacco.

The berries contain a substance that can curdle milk, and this quality has seen widespread application in the making of goat and other types of cheese at various pueblos. Sandia tribal elder Felipe Lauriano remembers his grandmother crushing the large, round berries in a metate, or grinding stone, when she wanted to make cheese. The use of horse-nettle berries as a substitute for commercial rennet seems to have been continued by a few traditional homemakers until very recently.

WILD POTATO
Tomato family
(*Solanum jamesii; S. fendleri*)

Two species of wild potato occur in our area. *Solanum jamesii* has white flowers and is more often associated with old fields in the Rio Grande region. Look for these plants along the Ruins Trail at the bottom of Frijoles Canyon at Bandelier. The flowers of *Solanum fendleri* are purple, and the plant is more likely to occur in Zuni country and farther south and west. Unlike the other *Solanum* (horse-nettle) featured in this book, the leaves of wild potatoes consist of a series of several smooth-margined, mostly paired leaflets, and the plants are not prickly.

Wild potatoes are, indeed, close relatives of the domestic potato, but their globe-shaped tubers are not much larger than grapes. However, they were regularly dug up with wooden sticks and boiled or baked at many pueblos. The tiny potatoes contain mildly toxic alkaloids, but the bitterness can be reduced by peeling and cooking.

SOLANUM JAMESII

The deliberate ingestion of clay with plants containing alkaloids is known to somewhat neutralize the toxicity. One ethnobotanist reported on this practice at Zuni just after the turn of the century: "The potato is eaten raw, and after every mouthful a bite of white clay is taken to counteract the unpleasant effect of the potato in the mouth" (Stevenson 1915).

Wild potatoes are most often found growing in cleared valley bottoms, some of which appear to be ancient, long-abandoned fields, or near old habitation sites. It seems that the tubers may intentionally have been divided and planted at some pueblos in times past, although this practice is not known today. Wild potatoes still can be found growing with other wild edible plants among the patches of corn, melons, and squashes in the old-style gardens still favored by a few Puebloans, among them David Chavez, but he says the potatoes in his garden west of Zuni Pueblo are too small to bother with nowadays. We have easily collected a pot of wild potatoes with a trowel in an old field in southwestern New Mexico. We washed and cooked them briefly. Seasoned with butter, salt, and pepper they were delicious, just like new potatoes.

PAINTBRUSH
Snapdragon family
(*Castilleja integra* and similar species)

Called paintbrush because each plant resembles a cluster of scarlet-dipped artist's brushes, this perennial grows about a foot high and has linear, inch-long leaves that are smooth on top and fuzzy on the undersides. The sticky floral bracts making up the flower are bright red with a little green at the base. The ubiquitous paintbrush brightens grasslands as well as piñon-juniper and ponderosa forests. It can be seen blooming all summer long at Bandelier and along most northern New Mexico roads that pass though the lower foothills and mountains.

Paintbrush has become a favorite native plant for wildflower gardens ever since someone discovered it partially acts as a parasite on grama grass roots and thrives best if the two species are planted together.

CASTILLEJA INTEGRA

Surprisingly, no evidence of prehistoric use of this plant has yet turned up in excavations; only contemporary uses are known. The most common historic use for paintbrush has been in conjunction with other plants to color deerskins red. It also was an ingredient, along with minerals, in the production of black paint. At Jemez it was said that mixing the flower bracts with chile seeds would prevent spoilage of the latter. The Tewa prepared a bathing solution from the whole plant, the purpose of which was to alleviate the aches and pains that inevitably attend long outdoor ceremonies.

The Tewa also ate raw paintbrush blossoms. People from Tesuque report that the flower parts are still sucked for their nectar, much in the same way that older readers may have sucked honeysuckle and other blossoms in their youth.

BUFFALO GOURD

Gourd family
(*Cucurbita foetidissima*)

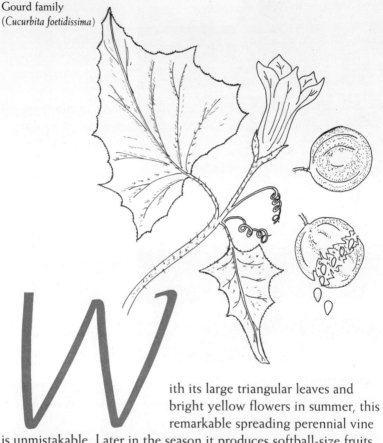

With its large triangular leaves and bright yellow flowers in summer, this remarkable spreading perennial vine is unmistakable. Later in the season it produces softball-size fruits, which turn from striped green to golden as summer progresses. Finally, the vine shrivels and dies, leaving bleached gourds on the ground and below the surface a man-size fleshy taproot that may weigh up to 150 pounds after several years of growth.

Buffalo gourd grows in disturbed sandy places such as along rural roadsides and arroyos, mainly in the juniper-grassland ecozone and up into the lower piñon-juniper ecozone. It often grows near fences, where it gets some protection from trampling and a bit of extra water. Look for buffalo gourd at Jemez State Monument.

CUCURBITA FOETIDISSIMA

One early account suggests that the ripe gourds were used as rattles in Pueblo dances, but a craftswoman at Tesuque Pueblo disagreed, saying the skin is too thin and the gourd doesn't dry properly. Our own experiments seem to confirm her observations. Rather, it's the bottle gourd (*Lagenaria siceraria* sp.), introduced into the prehistoric Southwest from Mesoamerica at an early date, that was and still is the main implement for ceremonial rattles.

Buffalo gourd seeds have been recovered from Anasazi and even more ancient archaeological sites. Both the seeds and the blossoms probably were eaten in early days but definitely *not* the other plant parts, for they contain *cucurbitacins*, foul-smelling chemicals and perhaps the bitterest natural substances known to mankind.

Besides being noxious to humans and livestock, cucurbitacins have the peculiar quality of compulsively attracting certain species of insects and repelling others. Pueblo Indians along the Rio Grande seem to have learned of the latter effect early on, for several Puebloans have related accounts of using buffalo gourd insect repellant. A Cochiti elder told us he used to crush the gourd in water and sprinkle the liquid on squash plants in his garden to repel squash bugs. At Sandia tribal elder Felipe Lauriano recalls that in his youth the gourd and leaves were cut and placed in all four corners of pueblo homes to keep out insects. The ground roots once were used in sleeping quarters at Santo Domingo to get rid of bedbugs.

Buffalo gourd has been used for human medicine, too. Not many years ago Isleta Puebloans boiled the roots to extract a liquid used in treating chest pains, and the roots ground fine and mixed with water served as a laxative for Tewa people. At Zuni the seeds and flowers have been mixed with saliva to reduce swellings.

The gourds also contain saponin, an agent that produces soapy lather. Thus, they have been used for cleaning purposes over the years by many American Indians and other rural desert-dwelling people throughout the Southwest. At Sandia Pueblo gourds chunks were once rubbed on clothes as a kind of soap, and at Cochiti the spheres were cut in half for scrubbing pots and dishes. It's likely that at one time or another people from all the Rio Grande pueblos found ways to use the root or gourd of this plant as a useful detergent.

COCKLEBUR
Sunflower family
(*Xanthium strumarium*)

Cocklebur is best identified by its mid- to late-summer fruit, an egg-shaped, inch-long bur covered with hooked or straight spines. In spite of being an annual, cocklebur can get up to three feet tall. The burs easily attach to and are transported by wildlife, livestock, and pets, so cocklebur is likely to crop up on almost any disturbed ground in our area. Although not recorded along any of the trails covered here, it is common along the highway to Bandelier, between the Rio Grande crossing at Otowi Bridge and the town of White Rock.

According to an early account of Zuni Pueblo, "the seeds are ground, mixed with corn meal, made into cakes or balls, and steamed. This was a common dish among the poorer class of the Zuni in 1879" (Stevenson 1915).

XANTHIUM STRUMARIUM

But cocklebur seeds and foliage are much more often linked with medicine. At Zuni the seeds, which have been shown to contain a chemical with germicidal qualities, were also ground and applied to wounds, and sap from plant stems was once used in this way at Jemez. Other puebloans, including the Acoma, Laguna, and Santo Domingo, used plant parts for treating sores on livestock. Both Tewa- and Towa-speaking Puebloans once used cocklebur parts for treating diarrhea, vomiting, and urinary disorders. Indeed, this weedy plant, which occurs around the world but whose origin is unknown, has been associated with medicinal uses by aboriginal people throughout North and South America and has even been used for herbal medicine in China.

GUMWEED
Sunflower family
(*Grindelia aphanactis*)

As its name suggests, gumweed is a sticky plant, with resin exuding from the leaves and stems. Its yellow flowers of mid to late summer fairly glisten with milky goo. The flower head of this particular species is ball-like and composed of disk flowers only; other species of gumweed have both ray and disk flowers in each composite head.

This many-branched plant doesn't grow much more than eighteen inches tall and seems to be almost entirely restricted to wasteland—roadsides, overgrazed fields, and the like. A few gumweed plants grow in disturbed sites along the Frey and Ruins trails at Bandelier, and it thrives along the weedy trailside at Jemez State Monument.

As a group, gumweeds have a long history of pharmaceutical uses, both in the United States and Europe. Besides waxes and

GRINDELIA APHANACTIS

resins, gumweed parts seem to be loaded with acids and alkaloids that possess a variety of remedial effects. Puebloan uses during this century include a gumweed herbal tea drunk for kidney problems at Picuris and San Juan and for stomachaches at Acoma and Laguna. A liquid of dried and boiled parts was once used for cleaning skin abrasions at Jemez, and a paste of ground herbage was applied to sores by Picuris Puebloans. At Cochiti the sticky blossoms would be placed on an aching tooth.

During the archaeological salvage project in advance of the construction of Cochiti Dam, gumweed seeds were found preserved in a prehistoric bowl. Since they were mixed with highly palatable amaranth seeds, gumweed seeds may have had a dietary connection.

Elsewhere in the country gumweed has been used as a source for making yellow dye, but there is no record of any dye use by Puebloans in New Mexico.

YARROW
Sunflower family
(*Achillea lanulosa*)

Finely dissected fernlike leaves that are highly aromatic when crushed give away this fairly tall mountain-growing herb. It flowers profusely through most of the summer, with white, flat-topped clusters forming at the top of each stem.

Yarrow thrives where the ground is somewhat wet, such as along roadsides, where pavement runoff adds moisture, or in meadows of the ponderosa pine ecozone and higher. About the only places you'll see it from the trails we cover are on those passing through the riparian zone at the bottom of Frijoles Canyon or perhaps in the meadows around the Ponderosa Campground, although it's very common along the Ski Trail that traverses some of the higher country at Bandelier.

The genus name, *Achillea*, comes from Achilles, the famed Greek warrior, who supposedly used yarrow in staunching the flow

ACHILLEA LANULOSA

of blood from the wounds of his men. Thus, its pharmaceutical history goes well back in time. In fact, the volatile oils and acids contained in the leaves have known astringent qualities and also have been associated with treating colds and hemorrhoids and for hair and scalp care. These attributes are no doubt the reason that yarrow herbal teas and rinses are available to this day in many health food stores throughout the United States and Europe.

Most Indian peoples, including the Navajos and certainly the Pueblo tribes, have long known of the medicinal values of yarrow. Various Tewa-speaking Puebloans have chewed on or made a tea from the flowers, leaves, or roots to treat stomach disorders and toothaches. People from Zuni Pueblo know of the cooling effect possessed by the ground leaves of yarrow and have applied them to burns; at San Juan the leaves have been used in this way for sore lips. Other documented medicinal uses include the combating of chills at Cochiti and fever at Picuris and other northern pueblos.

It has been said that yarrow is eaten or used as flavoring at some pueblos, but it seems more likely that even when it is mixed with food, there is a medicinal reason for including this plant in the diet.

INDIAN TEA

Sunflower family
(*Thelesperma megapotamicum*)

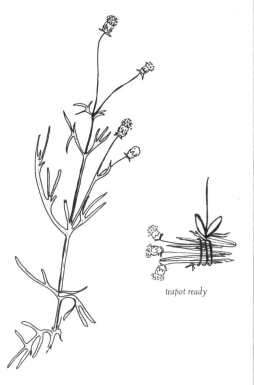

teapot ready

Indian tea, sometimes known as greenthread, is a slender plant growing up to two feet tall with sparse pairs of threadlike leaves attached to the stem. The distinctive golden flower heads that appear throughout the summer display no petals.

Common in the piñon-juniper ecozone throughout the Southwest, these plants often grow in patches on grassy flats and mesa tops. They are abundant in Frijoles Canyon at Bandelier but less common on the trail to Tsankawi and at Petroglyph National Monument.

As the name implies, the main use of this plant is for brewing a tea, and a delicious one at that. In our area virtually all the Pueblo tribes, as well as the Navajo and Apache, know this plant and consider it to be the best of several wild plant species for making tea.

THELESPERMA MEGAPOTAMICUM

Next to piñon nuts, Indian tea probably is the most universally sought plant for the table by modern Puebloans.

The stems and flowers are tied in bundles, then hung to dry for use throughout the year, but especially in winter. When steeped in boiling water a vivid orange to amber-colored beverage results. The taste resembles some commercial green teas and is very reminiscent of a popular tea made from the bark of the native sassafras tree in the eastern United States. As one Puebloan told us, it's a great natural, free alternative to Lipton.

Indian tea acts as a mild diuretic and seems to have other medicinal attributes. It has been used by Tewa and Santo Domingo Puebloans for stomach disorders and muscle cramps and by the Cochiti Indians for colds "instead of aspirin."

The Hopi, Navajo, and, possibly, Zuni have combined parts of this plant with several other wild species to manufacture a yellow or gold dye, but there is no record for dye use among Rio Grande Puebloans.

ANNUAL SUNFLOWER

Sunflower family
(*Helianthus annuus*)

The sunflower is the quintessence of the designer flower. When you look closely it is a bit more complicated, in that numerous yellow ray flowers, commonly thought to be petals, circle a dark cluster of disk flowers that have no petals. The sturdy stalks and spade-shaped leaves are covered with stiff hairs. Most species of sunflowers thrive in open, disturbed areas where there is a little extra water available. Roadsides and abandoned gardens and fields are ideal habitats. If you are on your way to any of the parks featured in this book in August or September, you won't miss them.

The domesticated sunflower apparently is the only aboriginally cultivated plant to originate north of Mexico. But archaeological remains are nearly nonexistent here, so the domestication probably did not happen anywhere in the New Mexico Pueblo Province. However, sunflowers seem to have become camp-following weeds,

HELIANTHUS ANNUUS

spreading from one village to the next up and down the Rio Grande, and then carried out onto the Great Plains during trading forays. This was where western and eastern varieties met and selection and domestication began.

In any case, wild sunflower plants are handy to Puebloans today and still have a variety of uses. For instance, San Ildefonso women put the flowers on their dresses for decoration during corn dances. Ramos Oyenque has sunflowers growing throughout his garden at San Juan Pueblo. The flowers are worn in women's hair there and woven into a circle to fit the hats worn by men during the harvest dance.

San Juan Pueblo men once used the light, strong stems for bird snares and arrows. The stalks are easily hollowed out and made into Indian flutes. At Jemez the stem juice of sunflowers is used for cuts, as is that of its weedy neighbor, the cocklebur. This is probably a good way to clean small wounds. At Zuni the root has been associated with treating rattlesnake bites.

We know through the direct evidence of coprolites and other archaeological finds in the Pueblo Province that sunflower seeds were an important seasonal food commodity—and they still are. From Phyllis Hughes's *Pueblo Indian Cookbook* we learn that "sunflower seeds, so rich a source of vitamins and minerals, have been widely used by Indians since prehistoric times, parched and eaten whole or ground into flour. These seeds, put through a grinder, may be mixed with flour or cornmeal for mush, bread making or soups.... Seeds of both wild and cultivated flowers are used."

7
ETHNOBOTANY IN NEW MEXICO

Terms and Techniques

*T*HIS BOOK COULD NOT HAVE BEEN WRITTEN if scientifically oriented people hadn't been studying ethnobotany a long time. The earliest application of the term *ethnobotany* is credited to John W. Harshberger, a professor at the University of Pennsylvania in 1896. He enthusiastically applied the term to the study of "plants used by primitive and aboriginal people" after he had identified the materials in baskets and sandals from an Anasazi cave site in Colorado. He did not use the words *primitive* and *aboriginal* in a pejorative sense, no more than the terms *primitive art* and *aboriginal artifacts*, or for that matter the word *antiques*, have negative connotation today. They were used to define ethnobotany as a study of the use of plants by indigenous peoples in a simple society. Later, of course, it was recognized that Puebloans and their ancestors had complex social systems; they did not passively employ what the environment offered but actually modified the plant world to suit their needs.

This modification of environment was sometimes intentional, as in the collecting and transporting of plants to eat or to transform into useful objects, incidentally depleting them in one area and sowing them in another. Or, it could be more subtle, as in the unknowing introduction of new weed species through the adoption and growing of weed-contaminated crop seeds ("dirty seed"). Manipulating the genes of certain species by selecting seed for color, size, or other perceived desirable traits, and growing them in separated garden plots so they would not hybridize, was certainly a conscious plant-breeding effort even if the Puebloans didn't define it as such in their own cultural terms.

The definition of ethnobotany has been qualified and broadened many times since 1896. For our purposes the roots of the word as defined in most dictionaries will suffice: *ethno* is the combining prefix signifying a human cultural group, race, or people; *botany* is a branch of biology dealing with plants. Ethnobotany is, therefore, the study of the mutual interactions of plants and human cultures or cultural groups.

The sources of ethnobotanical information have also broadened through the years. Folklore is now recognized as a means, both modern and ancient, of teaching *ethnoscience*. From wandering nomads to the more settled Puebloans, early southwestern peoples have been primarily concerned with obtaining food, shelter, and medicine. Their ancestors depended upon indigenous plants and animals for their basic needs. Having to apply themselves to learning about what was available to sustain life, they of necessity became applied ecologists. They passed down this information by example and through folktales. They watched the behavior of animals and saw what they ate. They learned through trial and error that many kinds of animals could be eaten but that the food these animals ate was not always safe. For instance, squirrels relish most mushrooms, and birds eat all sorts of berries with impunity, but some of those same mushrooms and berries can be poisonous to humans. So the people formed a special relationship with those animals whose dietary preferences and medicinal needs were similar to their own.

In this way the bear and the coyote became folk heroes. These two animals are omnivorous and use plants in similar ways; people could eat what both these animals ate. Even the animal and human medicinal needs were similar. Bears, for example, heal external injuries by rubbing their wounds on juniper bark; Puebloans also have used the bark of these trees to clean wounds. Unbeknownst to both bear and early Puebloan, juniper bark contains an effective antiseptic agent. By lending human character to these creatures in folk stories, the Puebloans seem to have taught techniques of survival, passing them down to succeeding generations.

The study of ethnobotany crosses over into various other disciplines. An understanding of Puebloan linguistics is essential to interpreting the puns and jokes of folk stories, folk taxonomy, and beliefs about the natural world. Using the language keeps a culture alive while revealing useful botanical information not available through other means.

Ethnologists observe and study a specific ethnic or part of an ethnic group's way of life, along with the things, material and traditional, that the individual or group uses in that society. If not trained in plant identification, the ethnologist will take specimens to a *botanist*, who identifies the plants and usually places *voucher specimens* in a university herbarium as a permanent physical archive of plant data, a crucial link between folk knowledge and western science and an essential part of ethnobotanical studies.

The art of collecting voucher specimens has changed little since the Swedish botanist Carolus Linnaeus dispatched collectors to various parts of the New World in the eighteenth century. Whole specimens or significant plant parts—flowers, fruits, leaves, and roots—are placed in presses between drying paper in their typical growing positions and allowed to dry. After drying, the fragile plants are carefully packaged and transported to an herbarium for precise identification. (In the eighteenth century, the specimens were frequently lost during transport; sometimes even their collectors vanished.) Much information is needed to render specimens scientifically useful. Notebooks are kept in which essential data are recorded: the specimen field number, date, location and habitat of collection, associated plants, and ephemeral characteristics of the specimen (for example, the color of the fresh flower). And, Eureka!, we occasionally find comments on the aboriginal use of the plant in these notebooks.

The greatest source of study material for the modern southwestern ethnobotanist has come from the field of *archaeology*. In the past, excavated botanical waste was routinely discarded, and only well-preserved artifacts were retained. Thus, the excavations undertaken about the turn of the century at places such as Frijoles Canyon yielded many museum-quality artifacts but little information about the aboriginal use of wild plants. Though cliff dwellings, caves, and rock shelters protected artifacts and dry vegetable materials from the elements and usually preserved them well, rarely have these places remained undisturbed.

Archaeological techniques have since become much more sophisticated. Open sites such as small pueblos or archaic pithouses are excavated nowadays for the scientific information they provide, not necessarily for museum-quality artifacts, which are seldom recovered anymore. Fill removed in the process of a dig is routinely sifted for botanical *macrofossils*—pieces large enough to be identified without a microscope. Within the zone of excavation great care is taken to collect as much potentially informative material as possible from the most promising settings for botanical evidence—places such as hearths, bins, niches, and storage rooms. Dirt from these excavated areas is saved and taken to the laboratory or worked on site. The dirt is mixed with water so delicate botanical material can float to the surface and be skimmed off, while the rest of the muddy water is poured through graduated sieves to catch microscopic seeds, plant-eating insects, or any other tiny particles that may furnish more information. The expertise of the ethnobotanist is further employed at this point in order to identify the microscopic plant remains.

Sometimes *paleoethnobotanists*, experts in the use of electron microscopes, are called upon to identify pollen, phytoliths (crystallized plant parts), and tiny diagnostic portions of seeds from prehistoric times. Paleoethnobotanist Glenna Dean recently discovered cotton (*Gossypium* sp.) pollen near Abiquiu, New Mexico, within the soil of some prehistoric gardens outlined in stone cobbles and covered with pebble mulch. This turned out to be not only the location the furthest north of any cotton-growing area but, at an elevation of slightly over six thousand feet, the highest known cotton site in the Southwest. Thus, over a thousand years ago prehistoric Puebloans must have developed a strain of cotton perfectly suited to a cooler climate and a shorter growing season and even more drought resistant than any previously suspected. This discovery also partially solved the mystery of what was grown in those strange little gardens perched atop the hills and mesas.

Charred seeds and other charred plant parts are usually considered to be the most reliable indicator that the material retrieved from a dig is *anthropogenic*, that is, brought to the site by humans, not buried later by wild animals such as rodents. Archaeologists presume that the deepest level of a site is the oldest and that all materials above it were deposited later. Therefore, any intrusion, such as a burrowing rodent or shoveling pot hunter, that mixes the materials can render the excavated remains less scientifically valuable, if not useless.

Dendrochronology, the science of dating wood by correlating tree rings, has been useful in dating prehistoric ruins as well as reconstructing climates. A cross section of a tree trunk has a series of rings that are counted in pairs and that represent one year's worth of growth. Usually there is a dark-colored narrow ring indicating slow trunk growth during winter and a light-colored wider ring indicating faster growth in spring and summer when more moisture is normally available. By matching the width patterns of rings to other trees, a sequence of climate had been worked out for many locations in the Southwest. For example, the very fine rings of trees growing on the Pajarito Plateau around A.D., 1250 indicate a severe drought lasting about ten years.

Other disciplines related to ethnobotany are often called upon to further interpret the material collected at an archaeological site. One such discipline is *palynology*, or the study of pollen, the nearly indestructible microscopic spores produced by the male parts of flowers. Pollen samples taken systematically from the various levels of an excavation and later identified can reveal which wind-pollinated plants were growing in the region during the time the structure was in use and, secondarily, the nature of the prevailing climate. For example, one-seed juniper pollen found in large amounts at a prehistoric site would probably indicate that a piñon-juniper woodland dominated the local environment at the time the site was occupied. Ecologists know that this type of woodland is associated with a temperate climate and roughly ten to fourteen inches of precipitation a year, barely enough for successful dry farming in good years.

The pollen from insect-pollinated plants is usually sticky and does not blow around. So a preponderance of this type, for instance, cholla pollen, would suggest that inhabitants of the site brought back parts of the cactus for their own use. Cholla pollen collected near what was determined to be an area where food had once been prepared provides strong evidence that the cactus buds were being eaten; in fact, we know from ethnographical literature and experience that such use continues to this day.

Botanical material sampled from prehistoric and historic sites may yield additional information. As measured by the carbon-14 dating method, charcoal in sufficient quantities often helps determine the age of a formerly occupied site. Comparative analysis of the chromosome structure (in plants such as amaranth) may indicate that new genetic material at some point entered the scenario. Simple morphological comparisons can also be rewarding. Comparison of corncobs from pre- and post-Spanish times reveals that, after the Spanish colonists entered New Mexico, bringing with them new strains of maize from Mexico, the average length of a native ear of corn in the area nearly doubled.

Recently the realization that commercial hybrid seeds produce plants that are seldom acclimated to arid lands and require excessive water, fertilizer, and pesticides has opened a whole new ethnobotanical search for "heirloom" seeds. Such seeds have been found among southwestern Indians who have selected and prized them for generations and are still growing plants from them today. Most notably, the organization Native Seeds/SEARCH has saved both ancient crop seed and seed from wild plants that are semidomesticated and provided them to gardeners and farmers of arid lands. Several of the Rio Grande Pueblo farmers have participated by saving and sharing seeds. Examples of the ancient crops now available are Santo Domingo melons; Acoma white dry-farmed corn; Isleta blue corn for heavy soils; Cochiti small red, yellow, brown, and white popcorn; and Zuni *tomatillos*.

Inadvertent Ethnobotany

THE FIRST RECORDED OBSERVATIONS of an ethnobotanical nature for wild plants in New Mexico were noted by a lost Spanish explorer; in the journal of the Portuguese leader of an aborted Spanish colony; and, nearly two hundred years later, in the report of Franciscan friars bent upon discovering a route to the garrison and town of Monterey on the California coast.

Cabeza de Baca, shipwrecked on the coast of Florida in 1527, walked through southern New Mexico on his way to civilization and much later wrote about his adventures. He mentions that some of the people he encountered were desert dwellers living in *rancherias*, temporary settlements. Sites near springs where stone potholes were used to beat piñon nuts from their cones can still be seen on his route. Prickly pear fruit, juniper berries, and ground whole dried herbs were the main plant foods. It is in this account that cooking in gourds by dropping in hot rocks is first mentioned.

Gaspar Castaño de Sosa's journal of his 1590 colonizing attempt records that Mexican Indians in his expedition party ground and ate mesquite beans they collected on the way to relieve the food shortages.

From the Franciscan friars of the 1776 Domingues-Escalante Expedition comes an early ethnobotanical note about threeleaf sumac: "There are also clumps of *lemita*, which is a red bead the size of the blackthorn's, and its coolness and taste very similar to the lemon's, so that in this country it is regarded as its substitute for making cool drinks." This and other tidbits are scattered throughout the early chronicles.

In New Mexico, ethnobotany was stimulated by commerce on the Santa Fe Trail. After the Mexican war for independence from Spain in 1821, New Mexico became part of Mexico, and trade with the United States was permitted, indeed encouraged, for the next twenty years.

William Gambel (1821–1849), a student of the Philadelphia Academy of Natural Science, is credited with having been the first botanist to collect specimens in New Mexico. A naturalist and a physician, he joined a group of trappers who in July 1841 took him on a transcontinental journey to Santa Fe, where he collected plants in the nearby mountains. Gambel oak (*Quercus gambelii*) is named for him, but he seems not to have collected any ethnobotanical information about the species. Some potentially useful ethnobotanical information is found in inadvertent comments recorded in the journals and letters of other travelers and explorers of this period, but these also lack specific mention of the Indian populations of New Mexico.

With the invasion of New Mexico by the United States in 1846 came several army officers trained in natural history. Traveling with them was Frederick A. Wislizenus, a physician-botanist from Germany who contributed to science the type specimen for our beloved piñon, though, to our knowledge, contributed no information about the use of its seed. Most of the journals, reports, and memoirs of this group included scraps of ethnobotanical information invaluable to this day since they were recorded nowhere else.

Lt. James W. Abert spoke with a Spanish schoolteacher about Indian bean gardens on the mesa tops. Other young officers described the carefully tended orchards at various pueblos and the peoples' generosity in sharing their fruit with the hot, thirsty soldiers. They told of Indians collecting wild cherries, plums, and threeleaf sumac. Less direct, but still relevant, information comes from these officers' observations of the New Mexican landscape and what was growing on it. Lt. W. H. Emory mentions a valley about fifteen miles from Santa Fe that had ample grama grass for horses and wild potatoes "large as a wren's egg." He also mentions adobe walls at the larger haciendas with cacti growing on top, and he notes that the common people did not have much in the way of gardens while those at the pueblos were lush and well kept.

After the Civil War numerous surveys were performed with an eye to American expansion. Among these studies, which routinely employed botanists to assess the vegetational potentials of the country, were the United States Geological, Boundary, and Railroad surveys.

Although botanists, naturalists, and anthropologists were traipsing around New Mexico soon after it became a territory of the United States in 1850, not until after the railroad was completed to Albuquerque nearly forty years later did ethnologists arrive to study Indian society. But once the scientists reached Albuquerque, it was back to foot, mule, or horse and buggy.

Adolph Bandelier (1840–1914) was born in Berne, Switzerland, and came to the United States as a child. After returning to Switzerland for studies at the University of Berne, Bandelier spent ten years in the Southwest, making his headquarters in Santa Fe for most of that time. He was the first explorer of El Rito de los Frijoles, now within the national monument named for him. Probably more than any other early Southwest enthusiast, Bandelier set the foundations and patterns for archaeological studies in New Mexico. His anthropological work was mostly with the Cochiti, the Puebloans who claim the ruins in Frijoles Canyon as their ancestral home. Indeed, Bandelier's novel *The Delight Makers*, published in 1890, is set in Frijoles Canyon and was the first work of fiction that attempted to convey a picture of the daily life of a prehistoric people. The fruits of his ethnobotanical work, his plant collections, were first sent to the physician-botanist George Engelmann, a founding member of the Missouri Botanical Garden in St. Louis. To this day these collections have still not been completely mined of their useful ethnobotanical information.

Frank Hamilton Cushing (1857-1900) attended Cornell University, leaving that school to serve as an assistant ethnologist on one of Maj. John Wesley Powell's expeditions to the Southwest. Powell, chief of the Smithsonian Institution's Bureau of Ethnology at that time, recognized Cushing's sometimes awkward genius for ethnography. Remarking upon Cushing's fragile nature, Powell also noted that the ethnologist had been freed by his family to find his own genius in the forests above his New York homestead, where "he found sermons in stone and books in running brooks." Cushing's first expedition was actually led by Col. James Stevenson for the purpose of studying the Zuni, taking notes, and purchasing Zuni crafts, tools, pottery, and household objects. When the expedition ended in 1879, Cushing decided to stay with the Zuni and learn their way of life. He actually moved into the Pueblo governor's house, becoming a fact of their life. They eventually decided to make a Zuni out of this inept but sympathetic stranger. Cushing spent five years with them then was called back to duty with the Smithsonian; his stay at Zuni resulted in the classic 1920 book, *Zuni Breadstuff*. No one informed the Zuni of Cushing's death in 1900. Although he was regarded as a mixed blessing by some, others mourned his absence from "home" for more than fifty years after he had left them.

Ethnobotany Comes of Age in New Mexico

PAUL CARPENTER STANDLEY (1884–1963) obtained his master of science degree at New Mexico State University in 1909. Standley's paper, "Some Useful Native Plants of New Mexico" (1912), was the first ethnobotanical work to come out of the newly minted state of New Mexico. Working from the National Herbarium, he and E. O. Wooton published *The Flora of New Mexico* in 1915.

Privately educated, Matilda Coxe Stevenson (1849–1915) accompanied her husband, Col. James Stevenson, on early trips to New Mexico, and upon his death in 1889, she became officially affiliated with the United States Bureau of American Ethnology. Mrs. Stevenson is noted for her studies and articles about the Taos, the Tewa, and especially the Zuni. Her early ethnological writings contain some botanical information, but her major work of present interest is *Ethnobotany of the Zuni Indians*, produced in 1915.

Botanist Wilfred William Robbins (1884-1952) worked in the Southwest while he was both a Ph.D. candidate at the University of Chicago *and* a professor of biology at the University of Colorado; Robbins's plant collection from the Rito de los Frijoles area still resides in the herbarium of the latter school. He and ethnologists John P. Harrington and Barbara Freire-Marreco published the landmark *Ethnobotany of the Tewa Indians* in 1916.

Edward F. Castetter (1896-1978) was professor of botany at the University of New Mexico from 1928 to 1961. He is known for having developed a unique master of science program in ethnobotany. His first student, Sarah Louise Cook, successfully completed an ethnobotanical master's thesis on the Jemez Indians in 1930. Castetter had a succession of students in ethnobotany, including Volney H. Jones (*The Ethnobotany of the Isleta Indians*, 1931), who later became an eminent ethnobotanist at the University of Michigan, and George R. Swank (*The Ethnobotany of the Acoma and Laguna Indians*, 1932).

By the 1930s, the collection of historic ethnobotanical information had begun to grow almost exponentially despite the limiting effects of World War II. Yet three decades later the focus of the discipline had already shifted to paleoethnobotany, the intensive study of archaeological materials. Another shift is evident in the attention given to folk medicine from the 1970s to the 1990s, perhaps owing to the momentum of the back-to-nature movement that had begun a decade earlier. Currently, ethnobotany and its connections with folklore are being viewed as teaching tools by Puebloan educators at the college level, as well as by their colleagues in primary and secondary schools. It is beyond the scope of this book to follow these many fascinating side streams of ethnobotany. We refer those who wish to further explore some of these subjects to the bibliography at the end of this book, especially the recent works by Richard I. Ford, Gary Paul Nabhan, and Vorsila I. Bohrer.

Suggested Reading

Adams, Karen R.
1980   Pollen, Parched Seeds and Prehistory: A Pilot Investigation of Prehistoric Plant Remains from Salmon Ruin, a Chacoan Pueblo in Northwestern New Mexico. *Eastern New Mexico Contributions in Anthropology*, Vol. 9. Eastern New Mexico Univ., Portales, NM.

Bohrer, Vorsila L.
1986   Guideposts in Ethnobotany. *Journal of Ethnobiology* 6(1):27–43.

Ewan, Joseph and Nasta Dunn
1981   *Biographical Dictionary of Rocky Mountain Naturalists, 1682–1932.* W. Junk Publishers, The Hague/Boston.

Ford, Richard I., ed.
1978   *The Nature and Status of Ethnobotany.* Anthropological Papers No. 67. Univ. of Michigan, Ann Arbor.

McKelvey, Susan Delano
1955   *Botanical Exploration of the Trans-Mississippi West, 1790–1850.* Arnold Arboretum of Harvard Univ., Jamaica Plain, MA.

8

RECENT MODIFICATIONS TO THE LANDSCAPE

*T*HE AVERAGE TEMPERATURE, precipitation, and seasonal conditions in the upper Rio Grande province haven't varied a great deal for several thousand years, yet the visual landscape and its ecology that parallel the meteorological conditions have been markedly transformed, especially in the past century or so. Changes in climate haven't been the primary factor in this transformation; human activity, of course, is the cause.

The most sweeping changes have occurred within the inner corridor of the Rio Grande, where cities such as Albuquerque, Santa Fe, and Española have mushroomed. Small farms and highly managed pastureland have replaced native vegetation in most of the vast rich floodplain between centers of human population.

Historically, a dense cottonwood *bosque* (Spanish for "woods" or "forest") covered much of the Rio Grande floodplain, from the northern pueblos to Isleta and south to the Mexican border. Over time the larger trees were selectively cut for fuelwood and other uses, then portions of the entire *bosque* were cleared for agriculture and homes. In recent years pumping has lowered water tables, and more of the *bosque* has been lost owing to the effects of irrigation and flood-control projects.

Human manipulation—including regulation of water flow from upstream dams, construction of human-made levees and drainage ditches, and the channelization of most sections of the main river, with artificial riprap replacing natural silt and sandy banks in many places—has been highly detrimental to the Rio Grande *bosque*. Virtual elimination of natural periodic flooding has meant that the cottonwoods and willows, which need flooded soils for germination of their seeds, seldom reproduce anymore. These disturbances to the natural regime have encouraged exotic trees such as salt cedar, Russian olive, and Siberian elm to invade the *bosque* and its perimeter. The *bosque*, where it still exists, is now a narrow band of fragmented woodland—a, few remaining beads of native cottonwoods and willows along the thread of the once-great river. Remnants of the *bosque* have been best preserved where the river runs through the seven pueblos that border it and in a few public reservations, such as the Corrales Bosque Preserve, an eight-mile-long stretch of mature riparian woodland adjacent to the village of Corrales.

Changes to the natural landscape are not as obvious beyond the highly developed Rio Grande corridor, but, nonetheless, they have been substantial. Grazing and trampling by domestic livestock have probably had the greatest impact over the years. Beginning in the mid-1700s, Spanish settlers grazed sheep, goats, and cattle on a year-round basis throughout much of the region. The grazing of livestock in the territory of New Mexico peaked in the 1880s, when, it

SHEEP LAYING WASTE TO A MEADOW IN THE JEMEZ MOUNTAINS, CA. 1935. GRAZING AND TRAMPLING BY DOMESTIC LIVESTOCK PROBABLY HAS HAD THE GREATEST DETRIMENTAL IMPACT TO THE LANDSCAPE OF ANY FORCE DURING THE PAST TWO CENTURIES.

is estimated, nearly four million sheep, more than one million cattle, and countless horses—mostly owned by a handful of powerful Anglo barons—devoured the rangelands.

Although the numbers of cattle have been greatly reduced since then, and sheep or goat grazing is no longer much of a factor in this part of the state, the long-term effects of past and present overgrazing aren't easily reversed. Some highly palatable cool-season grasses, such as Indian ricegrass and needlegrass (*Stipa* spp.), both of which are thought to have once provided important grain foods for the native people, may have been brought to near extinction locally and never have fully recovered. Broom snakeweed, cholla cactus, and other opportunistic shrubs, as well as oneseed juniper, have invaded former grasslands. In many places on the Pajarito Plateau the thin topsoil that was denuded of its grass cover has been stripped away by periodic violent summer downpours. In the Rio Puerco drainage a few miles south and west of the Jemez Mountains, gullying and sheet erosion became so widespread that this river gained national notoriety earlier in the century for having carried the densest-known load of suspended sediments of any stream in the nation.

Overgrazing created another side effect: the reduction of surface grass and brushy fuels that allows wildfires to spread, particularly in the piñon-juniper and ponderosa pine ecozones. Reduced fuel loads soon resulted in fewer natural wildfires throughout the region. At one time the natural frequency of wildfires in any single location on the Pajarito Plateau had been every five to fifteen years. These were primarily low-intensity surface fires, burning the grasses, herbs, and litter of open forest understories. A sharp drop in fire frequency occurred in the late 1880s, when grazing peaked, and then gradually dropped off even further into the period, decades later, when human fire control efforts became highly organized and more sophisticated.

Ecologists now understand that for many ecosystems frequent low-intensity blazes are an essential ingredient in the development and maintenance of natural ecological conditions. For most vegetation types of northern New Mexico, all-out fire suppression is disruptive. More than one hundred years of trying to suppress every local wildfire in ponderosa pine forests has allowed an unnatural accumulation of needle litter and woody fuels and a greater-than-natural density of young trees. This change in forest structure and the provision of additional fuel has meant that the few fires that escape control and spread tend to build up a much greater heat intensity, leading to the potential for holocaustic fires. On the Pajarito Plateau the classic example of this phenomenon was the La Mesa Fire of 1977, which burned more than 15,000 acres in the heart of the ponderosa pine ecozone. In doing so, it converted forest to open grassy shrubland and threatened numerous archaeological sites within Bandelier National Monument, although, fortunately, none were seriously damaged.

Today the federal land management agencies are taking a more ecologically enlightened approach. Deliberately igniting fires under carefully planned prescribed conditions

THE LA MESA FIRE OF 1977 CONVERTED A PONDEROSA PINE FOREST TO A GRASSY SHRUBLAND.

and specific circumstances, they are also allowing even some lightning-ignited fires to burn, as long as such fires can be contained within predetermined boundaries and not threaten habitations or private land.

Beginning in the late nineteenth century and continuing right up to the present, commercial logging, at first for the production of railroad ties and scaffolding for mine operations and later for saw timber, has taken a huge toll on the old-growth forests of the Jemez, Sangre de Cristo, and other ranges bordering the Rio Grande. For several decades after World War II the practice of chaining or otherwise mechanically ripping out junipers and piñon pines on native woodland became national policy for the land-managing agencies. Thousands of acres of the Carson, Santa Fe, and Cibola national forests, as well as other public lands, were denuded of trees. In theory this terrain was to be converted to more "productive" grasslands for grazing additional livestock, but in reality this practice is seldom cost efficient or ecologically sensible. Among other consequences, food and cover for wildlife are likely to diminish.

Centuries of cutting piñon trees for fuelwood has also left its mark. In some areas, past culling of mature piñons from piñon-juniper woodlands has resulted in a shift strongly favoring the junipers and a scarcity of magnificent old piñons. Piñons may well be the slowest-growing trees in the region. It takes an average of 180 years for a piñon to reach a one-foot diameter and up to 500 years to produce an old tree. Thus, the survival of mature specimens may be in jeopardy, and a growing number believe that the state tree of New Mexico deserves more official protection than it now receives.

The list of modifications or outright intrusions to the natural landscape seems endless. Early in the twentieth century escaped domestic burros multiplied and spread across the Pajarito Plateau, creating havoc with natural vegetation until they were finally brought under control in 1983. Even wild animals can get out of hand. Most recently, herds of elk, which are protected from hunting within the boundaries of Bandelier National Monument and on the lands of Los Alamos National Laboratory, have so increased as to cause substantial damage to some of the native meadowlands. Culturally important plants such as banana yucca—which can literally have its heart eaten out by a hungry elk—have been severely overbrowsed, and trampling has had its impact on archaeological sites.

Water impoundments have devoured huge bites of land, such as at Cochiti Dam and Reservoir on the Rio Grande (eight square miles of bottomland affected) and Abiquiu Dam on the Rio Chama (nine square miles inundated). The proliferation of paved highways and power line rights-of-way, to say nothing of urban expansion—for example, the industrial and residential complex at Los Alamos, with its satellite bedroom community of White Rock, all in the heart of the Pajarito—create barriers that impede the movement of wildlife and introduce other environmental disturbances that severely fragment natural vegetative communities. Fragmentation of this sort nearly always results in unfortunate long-term effects on the ecological health of the land.

Of course, most of these changes to the landscape have been in the name of "progress." Who can argue with the forward march of civilization? Actually, many can and do, knowing that, among other negatives, the cumulative impact of civilization has severely imperiled the traditional way of life of the original inhabitants of this land. Every single one of the effects outlined above has had some direct or indirect consequence to Puebloan traditions, nearly always adverse. More than a few question whether the trade-off has ever been a viable one.

On a final note of optimism, however, a great deal of land, especially in the parks, forests, and pueblos at higher elevations, still remains in a condition approaching pristine. Fortunately, nature has the capacity for enormous resiliency, and wise land stewardship may still prevail.

Suggested Reading

deBuys, William
1985 *Enchantment and Exploitation.* Univ. of New Mexico Press, Albuquerque.

Rothman, Hal
1992 *On Ruins and Ridges—The Los Alamos Area Since 1880.* Univ. of Nebraska Press, Omaha.

Williams, Jerry L.
1986 *New Mexico in Maps.* Univ. of New Mexico Press, Albuquerque.

9
OTHER PLACES TO VISIT

So far we've focused on the trails at four different parks in New Mexico as being ideal places to find native plants, but there are many other parks in the region with excellent interpretive trails, as well as museums whose programs complement our story of wild plants and their uses by native peoples of Puebloan ancestry. Some of these places are briefly described below, starting with institutions in Albuquerque and proceeding farther afield. Even the most distant is within an easy day's drive from Albuquerque. Included are parks that feature a combination of interpretive centers and walking trails, as well as several museums of interest to anyone who cares to learn more about the subject. To aid in acquiring up-to-the-minute information, a contact address for each park or museum follows its description.

Indian Pueblo Cultural Center Not far from Old Town in Albuquerque, this center is a must for anyone wishing to learn about Pueblo Indians. Owned and operated by the nineteen pueblos of New Mexico, the showcase museum spotlights each contemporary pueblo, its culture, and its crafts and tells of seven centuries of American Indian history from the perspective of the Puebloans themselves. Traditional dances are performed throughout the year in the large central plaza, and films on various aspects of Puebloan life are shown daily. The center also houses a restaurant featuring authentic Pueblo Indian cooking and a gift shop containing the largest collection of authentic Indian jewelry in the Southwest. Open daily all year except for major holidays; modest entrance fee for museum. 2401 Twelfth Street NW, Albuquerque, New Mexico 87102.

Maxwell Museum of Anthropology Located on the University of New Mexico campus in Albuquerque, the Maxwell offers permanent exhibits covering the evolution of humans worldwide, along with the prehistory of people in the Southwest. One absorbing exhibit, "Restructuring the Past," explains various techniques, such as pollen analysis and dendrochronology, that are used by modern archaeologists to interpret evidence of the past. The bookstore offers what may be the most complete selection of publications on Indians in the state. Open daily, but hours vary; free. University of New Mexico, Albuquerque, New Mexico 87131.

Museum of Indian Arts and Culture One of four museums in Santa Fe that is operated by the Museum of New Mexico, this museum displays and interprets a superb permanent collection of prehistoric and contemporary Puebloan and other Southwest Indian pottery and other arts. Temporary displays and traveling exhibits often deal with Puebloan plant uses and related subjects. Bookstore features Puebloan arts. Open daily except for major holidays; entrance fee. 708 Camino Lejo, Santa Fe, New Mexico 87504.

Salinas Pueblo Missions National Monument Salinas Pueblo Missions presents both Spanish colonial and Puebloan legacies. The three units of this national monument provide short trails to various ruins that offer information on some plants and plant use. Interpretive centers cover the story of the Puebloans who lived in the area for centuries but finally departed in the late 1600s, apparently because of severe drought-inflicted famine. Park headquarters is at Mountainair, roughly seventy miles by highway southeast of Albuquerque. Open all year; free. P.O. Box 496, Mountainair, New Mexico 87036.

Pecos National Historical Park The ruins of two mission churches that once served Pecos Pueblo are the main attractions of this park located twenty-five miles southeast of Santa Fe by way of Interstate 25. There is little to see of the mostly buried structures, last inhabited by a dwindling population in 1838. In that year they left their home and walked nearly a hundred miles west to join their fellow Towa-speaking kinspeople at Jemez Pueblo. The park features a self-guiding trail to the ruins where you can identify a number of wild plants once used. An excellent museum deals with much of the subject matter we've covered. Open all year; modest entrance fee. P.O. Drawer 418, Pecos, New Mexico 87552.

Ghost Ranch Living Museum Operated by Carson National Forest and located a few miles north of Abiquiu, this small indoor-outdoor museum is more a zoological park, but it does include a separate facility, "Gateway to the Past," that interprets the prehistory of the Chama River Valley, once inhabited by the ancestors of modern Tewa-speaking Puebloans. Exhibits describe their fascinating grid-garden agriculture. Open daily all year; free. Ghost Ranch Living Museum, Abiquiu, New Mexico 87510.

Pot Creek Cultural Center Eight miles south of Taos, just off State Highway 518, this site has been developed by the United States Forest Service for learning about the ancestors of the modern Taos and Picuris pueblos. A mile-long self-guiding trail loop leads to a partially reconstructed pueblo village with prehistoric farming terraces and gardens. The trail emphasizes on-site interpretation of some common plants and their uses by Puebloans. Open daily during snow-free months; free. Camino Real Ranger District, Carson National Forest, P.O. Box 68, Peñasco, New Mexico 87553.

Aztec Ruins National Monument During Anasazi times a pueblo-style town was built along the Animas River where it flows out onto the plains from the mountainous country to the north. Today a self-guiding trail leads through the main ruins of this ancient village. The Park Service visitor center is on the outskirts of the city of Aztec. Open daily: modest entrance fee. P.O. Box 640, Aztec, New Mexico 87410.

Chaco Culture National Historical Park Prehistoric Puebloan architecture reached a zenith at Chaco Canyon during a 200-year period starting in the early A.D. 900s. This national park in northwestern New Mexico can be reached via secondary roads leading west from State Highway 44. It's worth the extra effort to travel to this out-of-the-way park, not only to see the many well-interpreted ruins but also to learn about the early Chacoans at the modern park museum or by taking a ranger-conducted walk along one of many trails here. Plants and their prehistoric uses are often the topic at campfire talks given in summer. Plan to spend at least a day. Open all year, but unpaved roads can be unpassable during inclement weather; modest entrance fee. Star Route 4, Box 6500, Bloomfield, New Mexico 87413.

Salmon Ruin Museum and Heritage Park Salmon Ruin, on the banks of the San Juan River between Farmington and

Bloomfield, was one of the larger Anasazi outlier villages occupied by people of the Chacoan culture during the eleventh century. Besides the ruins themselves, facilities here, operated by the San Juan County Museum Association, include an interpretive trail that features the identification and ethnobotanical uses of wild plants, as well as a small museum and an excellent bookstore. Open daily all year except for major holidays; modest entrance fee. P.O. Box 125, Bloomington, New Mexico 87413.

Mesa Verde National Park Established in 1906, Mesa Verde is probably the best-known and best-developed archaeological park in the country. Located in southwestern Colorado between Durango and Cortez, this park preserves thousands of prehistoric ruins spanning nearly eight hundred years during the Anasazi era. Several short self-guiding trails incorporate wild plants and their ancient uses into the interpretive story, and the park also provides excellent museum exhibits and a varied selection of ranger-guided walks and talks. Plan to spend at least a full day here. Open all year, but many facilities are closed in winter; entrance fee for vehicles. Mesa Verde National Park, Colorado 81330.

Anasazi Heritage Center Located just eighteen miles by highway from Mesa Verde National Park, the Anasazi Heritage Center, operated by the U.S. Bureau of Land Management, emphasizes hands-on exhibits that involve kids and grown-ups alike. It takes a modern approach to teaching how Puebloan ancestors who lived in the Four Corners region coped with their environment. Museum exhibits also explain some of the most recent techniques, some highly sophisticated, employed by archaeologists. A short self-guiding trail leads to a nearby ruin. Open daily except for major holidays; free. 27501 Highway 184, Dolores, Colorado 81323.

AN ANNOTATED LIST OF USEFUL PLANTS

The following list of useful wild plants is a compilation of all plant species occurring in the New Mexico Pueblo Province known to have been used by past or present Puebloans. Nomenclature normally follows the authority of Martin and Hutchins 1980. The list is based on a comprehensive review of the technical literature as well as information collected personally by the authors. Domesticated or recently introduced plants and uses that are principally ritualistic are not included.

Plant uses are subdivided into seven general categories. The users are classified under the five different Puebloan languages spoken in New Mexico pueblos today and further by geographical division for the Tiwa and Keres language groups. Prehistoric uses are derived from archaeological evidence, published or conveyed directly to the authors by contemporary authorities. The references cited in this list comprise only a fraction of the literature that was consulted.

ABBREVIATIONS USED:
(*) Apparently nonmedicinal use;
PH? Prehistoric use uncertain.

KEY TO CHART
Pueblo: **EK**-eastern Keres (Cochiti, San Felipe, Santo Domingo, Santa Ana, Zia); **JE**-Jemez; **MO**-used historically by most or all Pueblos; **NT**-northern Tiwa (Taos, Picuris); **PH**-prehistoric; **ST**-southern Tiwa (Sandia, Isleta); **TE**-Tewa (San Juan, Santa Clara, Nambé, San Ildefonso, Pojoaque, Tesuque); **WK**-western Keres (Acoma, Laguna); **ZU**-Zuni.

Reference: (See bibliography for full citation) **1**-personal knowledge by authors of this book; **2**-Bell and Castetter 1941; **3**-Bohrer 1960; **4**-Bohrer 1975; **5**-Brandt 1990; **6**-Camazine and Bye 1980; **7**-Castetter 1935; **8**-Clary 1984; **9**-Cook 1930; **10**-Ford 1992; **11**-Ford 1968; **12**-Foxx 1982; **13**-Gasser 1982; **14**-Hill 1982; **15**-Jones 1931; **16**-Kent 1983; **17**-Krenetsky 1964; **18**-Lang 1986; **19**-Lange 1959; **20**-Matthews 1989; **21**-Matthews 1992; **22**-Peckham 1990; **23**-Robbins et al. 1916; **24**-Stevenson 1912; **25**-Stevenson 1915; **26**-Stiger 1977; **27**-Swank 1932; **28**-Trierweiler 1987; **29**-Turney 1948; **30**-Underhill 1979; **31**-White 1942; **32**-White 1945; **33**-several authorities

	FOOD AND BEVERAGE (*)	MEDICINE	SMOKING OR CHEWING (*)	CONSTRUCTION	COLORING, TANNING, SOAP, ART, CRAFTS	CORDAGE, FIBER, FINE MATTING	IMPLEMENTS
Abies concolor (white fir)		NT, TE, WK 17, 23, 27		MO 33	MO 33		TE 23
Abronia fragrans (sand verbena)	WK 27	ZU 6					
Acer glabrum (Rocky Mountain maple)							ST 15
Acer negundo (box-elder)							PH, TE, WK 29, 23, 27
Achillea lanulosa (yarrow)	NT 17	NT, TE, EK, ZU 33					
Agastache pallidiflora (giant hyssop)	MO 7						
Aletes acaulis (stemless aletes)		WK 27					
Allium spp. (wild onion)	MO 33	TE, ST, ZU 10, 15, 1					
Alnus tenuifolia (alder)		WK 27			MO 33		NT 1
Amaranthus albus (tumbleweed)	NT 17						
Amaranthus graecizans (prostrate pigweed)	TE, EK, ZU 10, 7, 25						
Amaranthus powellii (amaranth)	ZU 3						
Amaranthus retroflexus (green pigweed)	MO 33						
Amaranthus spp. (amaranth)	PH, WK 33, 27	WK 27					
Ambrosia psilostachya (western ragweed)		TE, WK 10, 27					
Amelanchier sp. (serviceberry)	ST 15						
Amsonia brevifolia (amsonia)		ZU 25					
Andropogon scoparius (little bluestem)							TE 23
Androsace sp. (rock-jasmine)	ST 15						
Anemopsis californica (yerba mansa)		TE, ST, WK 10, 15, 27					
Apocynum cannabinum (Indian hemp)		WK 27					
Apocynum suksdorfii (dogbane)				ST 15			
Aquilegia elegantula (red columbine)		WK 27					
Arabis fendleri (Fendler rockcress)		WK 27					
Arctostaphylos pungens (pointleaf manzanita)	TE 10		EK 1				

Plant	Food and Beverage (*)	Medicine	Smoking or Chewing (*)	Construction	Coloring, Tanning, Soap, Art, Crafts	Cordage, Fiber, Fine Matting	Implements
Arctostaphylos uva-ursi (bearberry)			NT, TE, JE 17, 10, 9				
Aristida divaricata (poverty three-awn)		WK 27					
Artemisia campestris (sagebrush)		TE 23					
Artemisia carruthii (sagebrush)		ZU 25					
Artemisia filifolia (sand sagebrush)		TE, EK 23, 1					
Artemisia frigida (fringed sagebrush)		TE, ST, ZU 23, 15, 25					
Artemisia ludoviciana (Louisiana wormwood)		TE, WK 1, 27					
Artemisia tridentata (big sagebrush)	PH? 26	PH?, MO 26, 33					
Asclepias asperula (antelope horns)		TE 10					
Asclepias involucrata (dwarf milkweed)		TE, WK 24, 27					
Asclepias latifolia (broad-leaved milkweed)	NT, TE 17, 10	TE, ST, JE 10, 15, 1					
Asclepias spp. (milkweed)	MO 33			ST, WK, ZU 1, 27, 1		PH 16	
Aster commutatus (aster)	NT 17						
Aster hesperius (marsh aster)		ZU 25					
Astragalus amphioxys (crescent milkvetch)		ZU 6					
Astragalus lentiginosus (beakpod milkvetch)	JE, ZU 7, 25						
Astragalus spp. (milkvetch)	MO 7	JE 9					
Atriplex argentea (silverscale)	ST, WK 7, 27	ZU 6					
Atriplex canescens (fourwing saltbush)	PH?, EK 26, 19	PH?, JE, ZU 26, 9, 6		PH 33			EK, ST, JE 1, 15, 1
Atriplex powellii (ribscale)	EK, WK, ZU 7, 7, 25						
Baccharis sp. (baccharis)		WK 27					
Bahia dissecta (yellow ragweed)		WK, ZU 27, 25					
Bahia woodhousei (Woodhouse bahia)		TE, ZU 24, 6					
Baileya multiradiata (desert-marigold)				JE 9			
Berberis fendleri (Colorado barberry)	JE 9						
Berberis repens (Oregon grape)		NT, WK 17, 27			ZU 25		

	Food and Beverage (*)	Medicine	Smoking or Chewing (*)	Construction	Coloring, Tanning, Soap, Art, Crafts	Cordage, Fiber, Fine Matting	Implements
Berlandiera lyrata (green eyes)	WK 7						
Berula erecta (water-parsnip)		ZU 6					
Betula occidentalis (western water-birch)					JE 9		
Bouteloua curtipendula (side-oats grama)							TE, EK 23, 19
Bromus marginatus (mountain brome)							WK 27
Calliandra humilis (false mesquite)		ZU 6					
Calochortus gunnisonii (Mariposa lily)		WK 27					
Campanula parryi (Parry's bellflower)		ZU 25					
Castilleja integra (paintbrush)	TE, JE 1, 9	TE 24			ZU 25		
Ceanothus fendleri (buckbrush)	WK 27						
Celtis reticulata (netleaf hackberry)	PH, MO 33, 33						TE 16
Cercocarpus montanus (mountain-mahogany)		TE, WK 23, 27		PH 33	MO 33		PH, MO 33, 33
Chamaesyce polycarpa (spurge)		ZU 25					
Chenopodium album (lamb's quarters)	MO 7						
Chenopodium ambrosioides (Mexican tea)	TE 10	TE 10					
Chenopodium graveolens (goosefoot)		ZU 25					
Chenopodium leptophyllum (goosefoot)	MO 7						
Chenopodium spp. (goosefoot)	PH, MO 33, 33	TE, WK 10, 27					
Chrysothamnus nauseosus (rabbitbrush)	PH?, EK 20, 7	PH?, MO 20, 33			TE, ZU 10, 25		ST, WK 15, 27
Cirsium ochrocentrum (Santa Fe thistle)		TE, ZU 10, 6					
Cirsium pallidum (yellow thistle)		WK 27					
Cleome serrulata (Rocky Mountain beeplant)	PH, MO 33, 33	TE 23			PH, MO 33, 33		
Conopholis mexicana (Mexican squawroot)		WK 27					
Cornus stolonifera (red-osier dogwood)						JE 9	
Coryphantha spp. (pincushion cactus)	TE 23						
Cowania stansburiana (cliffrose)						WK 27	

257

Species	Food and Beverage (*)	Medicine	Smoking or Chewing (*)	Construction	Coloring, Tanning, Soap, Art, Crafts	Cordage, Fiber, Fine Matting	Implements
Croton texensis (doveweed)		PH?, MO 5, 33					
Cryptantha crassisepala (plains hiddenflower)		ZU 25					
Cryptantha jamesii (James hiddenflower)		ZU 6					
Cryptantha sp. (hiddenflower)		TE 24					
Cucurbita foetidissima (buffalo gourd)	ST 15	PH?, MO 12, 33			EK, ST 1, 1		EK 1
Cycloloma atriplicifolia (winged pigweed)	PH, ZU 13, 25						
Cymopterus bulbosus (wafer parsnip)	EK 7	WK 27					
Cymopterus fendleri (wild celery)	MO 33	TE 14					
Cymopterus purpureus (wafer parsnip)	TE, ZU 10, 1						
Cyperus aristatus (flatsedge)	WK 7						
Dalea formosa (feather indigobush)		JE 1					
Dalea nana (dwarf indigobush)		WK 27					
Dalea scoparia (indigobush)		WK 27					
Datura meteloides (jimsonweed)		PH?, WK, ZU 12, 27, 25					
Descurainia pinnata (western tansy mustard)	MO 7				PH, JE 22, 1		
Descurainia spp. (tansy mustard)	PH 8				PH 23		
Dithyrea wislizenii (spectacle-pod)		WK, ZU 27, 6					
Dyssodia acerosa (dogweed)		ST 15	WK 27				
Dyssodia papposa (fetid marigold)		TE 10					
Echinocereus fendleri (hedgehog cactus)	EK 7						
Echinocereus triglochidiatus (claret cup)	ST 15						
Echinocereus spp. (hedgehog cactus)	PH, WK 33, 7						
Ephedra nevadensis (rough joint-fir)	ZU 25	ZU 25			MO 30		
Ephedra torreyana (Torrey joint-fir)		TE, ST, WK 10, 15, 27					
Ephedra spp. (joint-fir)	PH? 26	PH? 26					
Equisetum laevigatum (smooth horsetail)	EK, WK 7, 27						

	FOOD AND BEVERAGE (*)	MEDICINE	SMOKING OR CHEWING (*)	CONSTRUCTION	COLORING, TANNING, SOAP, ART, CRAFTS	CORDAGE, FIBER, FINE MATTING	IMPLEMENTS
Eriogonum alatum (winged wild buckwheat)		ZU 6					
Eriogonum fasciculatum (wild buckwheat)		ZU 25					
Eriogonum jamesii (antelope-sage)		NT, WK, ZU 17, 27, 6					
Eriogonum racemosum (redroot wild buckwheat)		JE 1					
Eriogonum spp. (wild buckwheat)	PH? 33	PH?, TE, WK 33, 10, 27					
Erodium cicutarium (alfilaria)		JE, ZU 9, 6					
Erysimum capitatum (western wallflower)		WK, ZU 27, 6			WK 27		
Eupatorium herbaceum (western throughwort)		ZU 25			TE 24		
Eupatorium sp. (throughwort)			EK 19				
Euphorbia albomarginata (rattlsnakeweed)		WK, ZU 27, 6					
Euphorbia heterophylla (painted-leaf spurge)		TE 24					
Euphorbia serpyllifolia (thymeleaf spurge)		TE, ZU 24, 25					
Euphorbia spp. (spurge)	PH?, EK 28, 1	PH?, EK, ST 28, 19, 1					
Eurotia lanata (winterfat)		TE, ZU 10, 25					
Fallugia paradoxa (Apache plume)		EK 1					MO 33
Forestiera neomexicana (New Mexico olive)	TE 10			JE 9			
Fragaria spp. (wild strawberry)	MO 33						
Franseria acanthicarpa (bursage)		ZU 6					
Gaura coccinea (scarlet gaura)		TE 24					
Gaura parviflora (small-flowered gaura)		WK, ZU 27, 6					
Geranium caespitosum (purple geranium)		WK 27			JE 9		
Geranium fremontii (Fremont geranium)		NT 17					
Glycyrrhiza lepidota (wild licorice)		MO 33					
Gnaphalium wrightii (everlasting)		WK 27					
Grindelia aphanactis (gumweed)		PH?, MO 11, 33					
Gutierrezia sarothrae (broom snakeweed)		PH?, MO 23, 33					

Species	Food and Beverage (*)	Medicine	Smoking or Chewing (*)	Construction	Coloring, Tanning, Soap, Art, Crafts	Cordage, Fiber, Fine Matting	Implements
Muhlenbergia pungens (sandhill muhly)							ZU 1
Nicotiana attenuata (coyote tobacco)		MO 33					
Nolina microcarpa (beargrass)	ST 15	ST 15		PH, MO 2, 33		ST 15	
Notholaena sp. (cloak fern)		TE 23					
Oenothera albicaulis (palestem evening-primrose)		WK, ZU 27, 25					
Oenothera caespitosa (stemless evening-primrose)		ST 15					
Oenothera coronopifolia (evening-primrose)		ZU 6					
Oenothera hookeri (Hooker's evening-primrose)		ZU 6					
Opuntia imbricata (cane cholla)	PH, MO 33, 33						WK 27
Opuntia whipplei (Whipple cholla)	ZU 25						
Opuntia spp. (prickly pear)	PH, MO 33, 33	TE 10					WK 27
Orobanche fasciculata (cancer root)		ZU 25					
Orobanche spp. (broomrape)		TE, JE 24, 1					
Oryzopsis hymenoides (Indian ricegrass)	PH, ZU 33, 25						
Osmorhiza obtusa (sweet cicely)		ST 15					
Oxybaphus linearis (desert four-o'clock)		ZU 6					
Oxybaphus nyctagineus (desert four-o'clock)			WK 27				
Panicum obtusum (vine mesquite)				ST 15			
Panicum spp. (panic grass)	PH 26						
Parryella filifolia (dune broom)						ZU 25	
Parthenocissus inserta (western five-leaved ivy)	ST 15			JE 9			
Pectis angustifolia (lemoncillo)	WK 7	WK 27					
Pectis papposa (chinchweed)	ZU 33	WK 27					
Penstemon ambiguus (pink plains penstemon)		WK 27					
Penstemon barbatus (scarlet penstemon)	TE 10	TE 23					
Petalostemum candidum (white prairie clover)	TE, EK, WK 23, 7, 27	WK 27					

Species	Food and Beverage (*)	Medicine	Smoking or Chewing (*)	Construction	Coloring, Tanning, Soap, Art, Crafts	Cordage, Fiber, Fine Matting	Implements
Petalostemum compactum (prairie clover)		ZU 6					
Phacelia corrugata (scorpionweed)		WK 27					
Phacelia neomexicana (NM scorpionweed)		ZU 6					
Phacelia sp. (scorpionweed)		TE 24					
Philadelphus microphyllus (mock-orange)	ST 15				TE 14		
Phoradendron juniperinum (juniper mistletoe)	WK 27	TE, WK, ZU 33, 27, 6					
Phragmites communis (common reed)	TE 10			PH, JE 33, 1			PH, TE, JE, ZU 1, 23, 23, 1
Physalis foetens (NM groundcherry)	MO 7						
Physalis hederaefolia (groundcherry)	TE, ZU 10, 25						
Physalis virginiana (Virginia groundcherry)	MO 33						
Physalis spp. (groundcherry)	PH 33						
Picea pungens (Colorado blue spruce)		WK 27					
Pinus edulis (piñon pine)	PH, MO 33, 33	TE, ST, WK, ZU 14, 1, 27, 6	EK 1	PH, MO 33, 33	MO 33		
Pinus ponderosa (ponderosa pine)	ZU 7	TE 10		PH, MO 33, 33			PH, MO 29, 33
Plantago major (plantain)	WK 7	NT, ST, WK 17, 15, 27					
Plantago purshii (woolly Indian-wheat)		WK, ZU 27, 6					
Poa fendleriana (muttongrass)	PH 4						
Polanisia trachysperma (clammyweed)	MO 7			ST 15	TE 23		
Polygonum lapathifolium (knotweed)		WK, ZU 27, 25					
Populus angustifolia (narrowleaf cottonwood)					EK 19		
Populus fremontii (Fremont Cottonwood)	TE, ST, JE 1, 7, 7	TE 10		MO 33	MO 33		
Populus tremuloides (aspen)		TE, ST 23, 15		TE 10	MO 33		
Populus spp. (cottonwood)				PH 33			PH 1
Portulaca oleracea (common purslane)	MO 33	WK 27					
Portulaca retusa (notchleaf purslane)	PH, EK 33, 7						

Species	Food and Beverage (*)	Medicine	Smoking or Chewing (*)	Construction	Coloring, Tanning, Soap, Art, Crafts	Cordage, Fiber, Fine Matting	Implements
Proboscidea sp. (unicorn-plant)				TE 23			
Prosopis glandulosa (mesquite)	ST, WK 7, 7	ST 15					ST 15
Prosopis pubescens (screwbean mesquite)	ST 15						
Prunus americana (wild plum)	NT, TE, ST 7, 10, 15	TE 10					
Prunus virginiana (chokecherry)	PH, MO 33, 33	WK 27					MO 33
Pseudocymopterus montanus (mountian parsley)	TE, NT 7, 15						
Pseudotsuga menziesii (Douglas-fir)		MO 33		PH 33			
Psilostrophe tagetina (woolly paperflower)		ZU 25			WK, ZU 27, 25		
Psoralea lanceolata (lemon scurfpea)	TE, ZU 10, 6						
Psoralea tenuiflora (wedgeleaf scurfpea)		ZU 25					
Ptelea trifoliata (hoptree)	EK 7						
Pterospora andromedea (pinedrops)		WK 27					
Quercus gambelii (Gambel oak)	MO 33	TE, EK, ST 10, 31, 15					MO 33
Quercus spp. (oak)	PH 33			PH 21			PH 29
Ranunculus inamoenus (buttercup)		WK 27					
Ratibida columnifera (prairie coneflower)	ZU 25	NT, TE, ZU 17, 24, 25					
Ratibida tagetes (green coneflower)		WK 27					
Rhus glabra (smooth sumac)			TE, EK 10, 19				
Rhus trilobata (threeleaf sumac)	PH, MO 26, 33	EK, WK 31, 27	TE 32		PH, MO 33, 33		MO 33
Ribes inebrians (wild currant)	MO 33						
Ribes inerme (gooseberry)	MO 33						
Ribes spp. (gooseberry or currant)	PH 26						
Robinia neomexicana (New Mexico locust)	JE 9						PH, TE, JE, WK 29, 10, 9, 27
Rorippa sinuata (yellow cress)		ZU 25					
Rosa woodsii (wild rose)		NT, TE, ST 17, 10, 15					
Rubus parviflorus (western thimbleberry)	ST 15						

	FOOD AND BEVERAGE (*)	MEDICINE	SMOKING OR CHEWING (*)	CONSTRUCTION	COLORING, TANNING, SOAP, ART, CRAFTS	CORDAGE, FIBER, FINE MATTING	IMPLEMENTS
Rudbeckia laciniata (coneflower)	EK 7						
Rumex crispus (curlyleaf dock)	TE, ST 7, 15	ZU 6					
Rumex hymenosepalus (canaigre)	EK, ST 7, 1				MO 30		
Rumex mexicanus (wild dock)	EK 7	TE, WK, ZU 10, 27, 25					
Salix spp. (willow)		ST, ZU 15, 6		PH 33	PH, MO 33, 33		
Salvia reflexa (Rocky Mountain sage)		EK 19					
Sarcobatus vermiculatus (greasewood)		WK 27					EK, JE 31, 1
Scirpus sp. (bulrush)	WK 27						
Senecio douglasii (threadleaf butterweed)		JE, WK 9, 27					
Senecio multicapitatus (groundsel)		ZU 25					
Silene laciniata (Mexican campion)		WK 27					
Smilacina racemosa (false Solomon's seal)	TE 23						
Solanum elaeagnifolium (horse-nettle)	EK, ST, ZU 7, 1, 25	MO 33	EK 19				
Solanum fenderi (Fendler wild potato)	MO 33						
Solanum jamesii (wild potato)	EK, ST, WK 32, 15, 27						
Solanum rostratum (buffalo bur)		ZU 25					
Solanum triflorum (cutleaf nightshade)	WK, ZU 7, 25						
Solidago canadensis (Canada goldenrod)		ZU 25					
Solidago spp. (goldenrod)		WK 27			WK 27		
Sophora nuttalliana (silky sophora)	EK, WK 7, 7						
Sphaeralcea spp. (globe-mallow)	PH? 33	PH?, MO 33, 33					
Sporobolus contractus (spike dropseed)				ZU 25			
Sporobolus spp. (dropseed)	PH 33						
Stanleya pinnata (desert plume)		ZU 25					
Stephanomeria tenuifolia (wire-lettuce)		ZU 25					
Stipa spp. (needlegrass)	PH 4						

	FOOD AND BEVERAGE (*)	MEDICINE	SMOKING OR CHEWING (*)	CONSTRUCTION	COLORING, TANNING, SOAP, ART, CRAFTS	CORDAGE, FIBER, FINE MATTING	IMPLEMENTS
Swertia radiata (deer's ears)		ST 15					
Taraxacum officinale (common dandelion)	TE 23	TE 23					
Thalictrum fendleri (Fendler meadow rue)		WK 27					
Thelesperma longipes (cota)		ST 15					
Thelesperma megapotamicum (Indian tea)	MO 33						
Thelypodium wrightii (Wright's mustard)	MO 7				TE 23		
Tradescantia occidentalis (western spiderwort)	WK 7						
Tripterocalyx wootonii (Wooton sand verbena)		ZU 6					
Typha angustifolia (narrow-leaved cattail)						PH 18	
Typha latifolia (broad-leaved cattail)	MO 33			ST 15			
Verbesina encelioides (crownbeard)		TE, ZU 10, 25		WK 27			
Vicia americana (American vetch)	EK, WK 7, 27						
Vitis arizonica (canyon grape)	MO 33						
Xanthium strumarium (cocklebur)	ZU 25	MO 33					
Yucca baccata (banana yucca)	MO 33			MO 33	MO 33		MO 33
Yucca glauca (narrowleaf yucca)	MO 33	TE, WK 23, 27		MO 33	PH, MO 18, 33		MO 33
Yucca spp. (yucca)	PH 33			PH 33	PH 33		
Zinnia grandiflora (Rocky Mountain zinnia)		WK, ZU 27, 25		WK 27			
TOTAL PLANTS EACH USE	**129**	**180**	**16**	**16**	**38**	**8**	**32**

BIBLIOGRAPHY

The following writings include both works cited in the text and a selection of additional materials researched in the preparation of this manuscript.

Abert, J. W.
1962   *Report of Lieut. J. W. Abert of His Examination of New Mexico in the Years 1846–47.* 30th Congress, Senate Executive Report No. 23. Facsimile of original document by Horn and Wallace Publishers, Albuquerque.

Alexander, H. G. and P. Reiter
1935   Report on the Excavation of Jemez Cave, New Mexico. *University of New Mexico Bulletin*, Monograph Series 1(3):5–67, Univ. of New Mexico, Albuquerque.

Allen, Craig Daniel
1989   *Changes in the Landscape of the Jemez Mountains, New Mexico.* Unpublished Ph.D. dissertation, Univ. of California, Berkeley.

Allen, Patricia Joan, ed.
1989   *Uses of Native Materials in Zuni Pueblo.* Unpublished Zuni High School English Composition essays, 1984–85, Zuni, NM.

Anderson, Edgar
1952   *Plants, Man and Life.* Little, Brown & Co., Boston.

Arrhenius, Olof
1963   Investigation of Soil from Old Indian Sites. *Ethnos* 2-4: 122–136.

Barry, Patricia
1990   *Bandelier National Monument.* Southwest Parks and Monuments Association, Tucson.

Bell, Willis H. and Edward F. Castetter
1941   Ethnobiological Studies in the American Southwest. VII. The Utilization of Yucca, Sotol, and Beargrass by the Aborigines in the American Southwest. *The University of New Mexico Bulletin*, Univ. of New Mexico, Albuquerque.

Berry, Michael S.
1982   *Time, Space and Transition in Anasazi Prehistory.* Univ. of Utah Press, Salt Lake City.

Binford, Lewis R.
1968   Post-Pleistocene Adaptations. In *New Perspectives in Archaeology*, ed. by S. R. Binford and L. R. Binford. Aldine, Chicago.

1980   Willow Smoke and Dog's Tails: Hunter-Gatherer Settlement Systems and Archaeological Site Formation. *American Antiquity* 45(1): 4–20.

Blair, Mary Ellen and Lawrence R. Blair
1986   *Margaret Tafoya: A Tewa Potter's Heritage and Legacy*. Schifler Publishing, West Chester, PA.

Bohrer, Vorsila L.
1960   Zuni Agriculture. *El Palacio* 67:181–202.

1975   The Prehistoric and Historic Role of the Cool-Season Grasses in the Southwest. *Economic Botany* 29:199–207.

1983   New Life from Ashes: The Tale of the Burnt Bush (*Rhus trilobata*). *Desert Plants* 5:122–124.

1986   The Ethnobotanical Pollen Record at Arroyo Hondo Pueblo. In *Food, Diet, and Population at Prehistoric Arroyo Hondo Pueblo, New Mexico* by Wilma Wetterstrom. School of American Research Press, Santa Fe.

Brandt, Carol B.
1990   *Analysis of Archaeobotanical Remains from Three Sites Near the Rio Grande Valley, Bernalillo County, New Mexico*. Zuni Archaeologiccal Program, Ethnobiological Technical Series No. 90-2, Zuni, NM.

1991   *The River's Edge Archaeobotanical Analysis: Patterns in Plant Refuse*. Zuni Archaeological Program, Ethnobiological Technical Series No. 91-2, Zuni, NM.

1993   The Breath of Spring of the Zuni Mesas. *Native Plant Society of New Mexico Newsletter* 18(3):9.

Brody, J. J.
1990   *The Anasazi: Ancient Indian People of the American Southwest*. Rizzoli, New York.

Brooks, R. R.
1972   *Geobotany and Biochemistry in Mineral Exploration*. Harper's Geoscience Series. Harper and Row, New York.

Cajete, Gregory
1993   An Enchanted Land: Spiritual Ecology and a Theology of Place. *Winds of Change* 8(2):50–53.

1994   *Look to the Mountain-An Ecology of Indigenous Education*. Kivaké Press, Durango, CO.

Calvin, Ross
1951 *Lieutenant Emory Reports* (a reprint of Lieutenant W. H. Emory's notes of a military reconnaissance). Univ. of New Mexico Press, Albuquerque.

Camazine, Scott and Robert Bye
1980 A Study of the Medical Ethnobotany of the Zuni Indians of New Mexico. *Journal of Ethnopharmacology* 2(4):365–388.

Cannon, Helen L. and W. H. Starrett
1956 Botanical Prospecting for Uranium on La Ventana Mesa, Sandoval County, New Mexico. In *Geological Survey Bulletin* 1009-M. U.S. Government Printing Office, Washington, D.C.

Castetter, Edward F.
1935 *Uncultivated Native Plants Used as Sources of Food.* Ethnobiological Studies in the American Southwest No. 1. Univ. of New Mexico, Albuquerque.

Clary, Karen H.
1984 Anasazi Diet and Subsistence as Revealed by Coprolites from Chaco Canyon. In *Recent Research on Chaco Prehistory*, ed. by W. J. Judge and J. Schelberg. Reports of the Chaco Center, No. 8. National Park Service, Albuquerque.

Cook, Sarah Louise
1930 *The Ethnobotany of Jemez Indians.* Unpublished M.A. thesis, Univ. of New Mexico, Albuquerque.

Cordell, Linda S.
1984 *Prehistory of the Southwest.* Academic Press, Orlando.

Curtin, L. S. M.
1947 *Healing Herbs of the Upper Rio Grande.* The Laboratory of Anthropology, Santa Fe.

Cushing, F. H.
1920 *Zuni Breadstuff.* Indian Notes and Monographs, Vol. 8, National Museum of the American Indian, Smithsonian Institution, Washington, D.C.

Darwin, Charles
1909 *The Voyage of the Beagle.* P. F. Collier & Son, New York.

Dittert, Alfred E., Jr. and Fred Plog
1980 *Generations in Clay.* Northland Press, Flagstaff.

Doebley, John F.
1984 "Seeds" of Wild Grasses: A Major Food of Southwestern Indians. *Economic Botany* 38:52–64.

Drager, Dwight L. and Richard W. Loose
1977 An Ecological Stratification of the Southern Pajarito Plateau. In *Archeological Investigations in Cochiti Reservoir, New Mexico, Vol. 1: A Survey of Regional Variability*, ed. by Jan V. Biella and Richard C. Chapman. Office of Contract Archeology, Univ. of New Mexico, Albuquerque.

Dutton, Bertha
1963 *Sun Father's Way: The Kiva Murals of Kuaua*. Univ. of New Mexico Press, Albuquerque.

Ellis, Florence Hawley
1970 Irrigation and Water Works in the Rio Grande Valley. Paper presented at the 1970 Pecos Conference, Symposium on Water Control, Santa Fe.

1987 The Long Lost "City" of San Gabriel del Yungue, Second Oldest European Settlement in the United States. In *When Cultures Meet*. Papers from the October 20, 1984, conference held at San Juan Pueblo, New Mexico. Sunstone Press, Santa Fe.

Fall, P. L.
1988 Vegetation Dynamics in the Southern Rocky Mountains: Late Pleistocene and Holocene Timberline Fluctuation. Unpublished Ph.D. dissertation, Univ. of Arizona, Tucson.

Fellah, Abdulmunam Mohamed
1990 The Nutritional Value and Toxic Properties of Buffalo Gourd (*Cucurbita foetidissima*) Plant. Unpublished Ph.D. dissertation, Univ. of Arizona, Tucson.

Ford, Richard I.
1968 Floral Remains. In *The Cochiti Dam Archeological Salvage Project, Part 1: Report on the 1963 Season*, ed. by C. H. Lange. Museum of New Mexico Research Records No. 6, Univ. of New Mexico, Albuquerque.

1975 Re-Excavation of Jemez Cave, New Mexico. *Awanyu* 3(3):13–27.

1981 Gardening and Farming Before A.D. 1000: Patterns of Prehistoric Cultivation North of Mexico. *Journal of Ethnobiology* 1(1):6–27.

1984 Ecological Consequences of Early Agriculture in the Southwest. In *Papers on the Archaeology of Black Mesa, Vol. 2*, ed. by Stephen Plog and Shirley Powell. Southern Illinois Univ. Press, Carbondale.

1985 The Processes of Plant Food Production in Prehistoric North America. In *Prehistoric Food Production in North America*, ed. by Richard I. Ford, pp. 1–18. Anthropological Papers, No. 75. Museum of Anthropology, Univ. of Michigan, Ann Arbor.

1987   The New Pueblo Economy. In *When Cultures Meet*. Papers from the October 20, 1984 conference held at San Juan Pueblo, New Mexico. Sunstone Press, Santa Fe.

1992   *An Ecological Analysis Involving the Population of San Juan Pueblo, New Mexico*. Garland Press, New York.

Foxx, Teralene S.
1982   Vegetative Study. In *Pool of Cochiti Lake, New Mexico*, ed. by Lyndi Hubbell and Diane Traylor. National Park Service, Southwest Cultural Resources Center, Santa Fe.

Foxx, Teralene S. and Gail D. Tierney
1984   *Status of the Flora of the Los Alamos National Environmental Research Park—A Historical Perspective.* LA-8050-NERP. Los Alamos National Laboratory, Los Alamos.

Gasser, Robert E.
1982   Anasazi Diet. In *The Coronado Project Archeological Investigations. The Specialist's Volume: Biocultural Analyses*, compiled by Robert E. Gasser. Coronado Series 4, Museum of Northern Arizona Research Paper 23, Flagstaff.

Gumerman, George J., ed.
1988   *The Anasazi in a Changing Environment.* Cambridge Univ. Press, Cambridge.

Hall, Stephen A.
1985   Quaternary Pollen Analysis and Vegetational History of the Southwest. In *Pollen Records of Late-Quaternary North American Sediments*, ed. by Vaughn Bryant and R. G. Hollaway. American Association of Stratigraphic Palynologists, Dallas.

Hammond, George P. and Agapito Rey
1953   *Oñate, Colonizer of New Mexico 1595–1628, Vol. 1 and 2.* Univ. of New Mexico Press, Albuquerque.

1966   *The Rediscovery of New Mexico 1580-1594.* Univ. of New Mexico Press, Albuquerque.

Harrington, H. D.
1967   *Edible Native Plants of the Rocky Mountains.* Univ. of New Mexico Press, Albuquerque.

Harris, Marvin
1985   *Good to Eat - Riddles of Food and Culture.* Simon and Schuster, New York.

Harshberger, John W.
1896   The Purpose of Ethnobotany. *American Antiquarian.* 17:33–81.

Heiser, Charles B., Jr.
1955   The Origins and Development of the Cultivated Sunflower. *The American Biology Teacher* 17(5):161–167.

Here, R. F. and Davis Griffiths
1907   The Tuna as a Food for Man. *New Mexico Agricultural Experiment Station Bulletin* No. 64, Las Cruces.

Hill, James N. and William N. Trierweiler
1986   *Prehistoric Responses to Food Stress on the Pajarito Plateau, New Mexico: Technical Report and Results for the Pajarito Archeological Research Project, 1977–1985.* Report submitted to the National Sciences Foundation, Washington, D.C.

Hill, W. W.
1982   *An Ethnology of Santa Clara Pueblo, New Mexico,* ed. by Charles H. Lange. Univ. of New Mexico Press, Albuquerque.

Hoard, Dorothy
1989   *A Guide to Bandelier National Monument.* Los Alamos Historical Society, Los Alamos.

Housley, L. K.
1974   *Opuntia imbricata Distribution of Old Jemez Indian Habitation Sites.* Unpublished master's thesis, Department of Biology, Univ. of New Mexico, Albuquerque.

Hughes, Phyllis
1972   *Pueblo Indian Cookbook.* Museum of New Mexico Press, Santa Fe.

Indian Pueblo Cultural Center
Undated   *Yesterday and Today: Seven Centuries of American Indian History at the Indian Pueblo Cultural Center.* Indian Pueblo Cultural Center, Albuquerque.

Irwin-Williams, Cynthia
1973   The Oshara Tradition: Origins of Anasazi Culture. *Eastern New Mexico University Contributions in Anthropology,* Vol. 5, No. 1, Portales.

Jacobs, B. F.
1989   *A Flora of Bandelier National Monument.* Unpublished report on file at Bandelier National Monument, NM.

Jenkins, Myra Ellen
1987   Oñate's Administration and the Pueblo Indians. In *When Cultures Meet.* Papers from the October 20, 1984, conference held at San Juan Pueblo, New Mexico. Sunstone Press, Santa Fe.

Johns, Timothy
1990   *With Bitter Herbs They Shall Eat It.* Univ. of Arizona Press, Tucson.

Jones, Volney H.
1931 *The Ethnobotany of the Isleta Indians*. Unpublished master's thesis, Univ. of New Mexico, Albuquerque.
1938 An Ancient Food Plant of the Southwest and Plateau Regions. *El Palacio* 44:41–53.
1942 A Native Southwestern Tea Plant. *El Palacio* 49(12):272–280.

Kent, Kate Peck
1983 *Prehistoric Textiles of the Southwest*. School of American Research, Santa Fe, and Univ. of New Mexico Press, Albuquerque.

Kidder, A. V.
1915 Pottery of the Pajarito Plateau and of Some Adjacent Regions in New Mexico. *Memoirs of the American Anthropological Association* 12:407–462.
1932 *The Artifacts of Pecos*. Yale Univ. Press, New Haven.

Krenetsky, John C.
1964 *Phytosocialogical Study of the Picuris Grant and Ethnobotanical Study of the Picuris Indians*. Unpublished M.A. thesis, Univ. of New Mexico, Albuquerque.

Lamb, Susan
1993 *Petroglyph National Monument*. Southwest Parks and Monuments Association, Tucson.

Lang, Richard W.
1986 Artifacts of Woody Materials from Arroyo Hondo Pueblo. In *Food, Diet, and Population at Prehistoric Arroyo Hondo Pueblo, New Mexico*, by Wilma Wetterstrom. School of American Research Press, Santa Fe.

Lange, Charles H.
1959 *Cochiti: A New Mexico Pueblo, Past and Present*. Univ. of Texas Press, Austin.

Lanner, Ronald M.
1981 *The Piñon Pine—A Natural and Cultural History*. Univ. of Nevada Press, Reno.

Malde, Harold E.
1964 Environment and Man in Arid America. *Science* 145:123–129.

Martin, Paul Schultz
1963 *The Last 10,000 Years: A Fossil Pollen Record of the American Southwest*. Univ. of Arizona Press, Tucson.

Martin, William C. and Charles R. Hutchins
1980 *A Flora of New Mexico*. J. Cramer, West Germany.

Matthews, Merideth H.
1989 Macrobotanical Analysis. In *Bandelier Archeological Excavation Project: Research and Design and Summer 1988 Sampling*, ed. by Timothy A. Kohler. Reports of Investigations 61, Washington State Univ., Pullman.

1992 Macrobotanical Analysis. In *Bandelier Archeological Excavation Project: Summer 1990 Excavations at Burnt Mesa Pueblo and Casa del Rio*, ed. by Timothy A. Kohler and Matthew J. Root. Reports of Investigations 64, Washington State Univ., Pullman.

Minge, Ward Alan
1979 Effectos del pais: A History of Weaving Along the Rio Grande. In *Spanish Textile Tradition of New Mexico and Colorado*, ed. by Nora Fisher. Museum of New Mexico Press, Santa Fe.

Minnis, Paul E.
1981 Seeds in Archeological Sites: Sources and Some Interpretive Problems. *American Antiquity* 46:143–152.

1985 Domesticating Plants and People in the Greater American Southwest. In *Prehistoric Food Production in North America*, ed. by Richard I. Ford. Anthropological Papers, No. 75. Museum of Anthropology, Univ. of Michigan, Ann Arbor.

1989 Prehistoric Diet in the Northern Southwest: Macroplant Remains from Four-Corners Feces. *American Antiquity* 54:543–563.

Mosimann, James E. and Paul S. Martin
1975 Simulating Overkill by Paleoindians. *American Scientist* 63:304–313.

Moore, Michael
1979 *Medicinal Plants of the Mountain West*. Museum of New Mexico Press, Santa Fe.

Nabhan, Gary Paul
1985 *Gathering the Desert*. Univ. of Arizona Press, Tucson.

National Research Council
1984 *Amaranth: Modern Prospects for an Ancient Crop*. National Academy Press, Washington, D. C.

Niethammer, Carolyn
1974 *American Indian Food and Lore*. MacMillan Publishing Co., New York.

Orcutt, Janet Dale
1991 Environmental Variability and Settlement Changes on the Pajarito Plateau, New Mexico. *American Antiquity* 56(2):315–332.

Peckham, Stewart L.
1990 *From This Earth: The Ancient Art of Pueblo Pottery*. Museum of New Mexico Press, Santa Fe.

Potter, Loren D.
1972 Plant Ecology of the Walakpa Bay Area, Alaska. *Journal of the Arctic Institute of North America* 25(2):115–130.

Potter, Loren D. and Richard Young
1983 Indicator Plants and Archaeological Sites, Chaco Canyon National Monument. COAS: *New Mexico Archaeology and History* 1(4):19–37.

Preucel, Robert Washington, Jr.
1988 *Seasonal Agricultural Circulation and Residential Mobility: A Prehistoric Example from the Pajarito Plateau, New Mexico*. Unpublished Ph.D. dissertation, Univ. of California, Los Angeles.

Reher, Charles A. and D. C. Witter
1977 Archaic Settlement and Vegetative Diversity. In *Settlement and Subsistence Along the Lower Chaco River: The CGP Survey*, ed. by Charles A. Reher. Univ. of New Mexico Press, Albuquerque.

Reinhart, Theodore Russell
1968 *Late Archaic Cultures of the Middle Rio Grande Valley, New Mexico: A Study of the Process of Culture Change*. Unpublished Ph.D. dissertation, Univ. of New Mexico, Albuquerque.

Robbins, W. W., J. P. Harrington, and Barbara Freire-Marreco
1916 *Ethnobotany of the Tewa Indians*. Bureau of American Ethnology Bulletin No. 55, Smithsonian Institution, Washington, D. C.

Rose, Martin R., Jeffrey S. Dean, and William J. Robinson
1981 *The Past Climate of Arroyo Hondo, New Mexico, Reconstructed from Tree Rings*. School of American Research Press, Santa Fe.

Rothman, Hal
1989 Cultural and Environmental Change on the Pajarito Plateau. *New Mexico Historical Review* 64:185–211.

Sando, Joe
1991 *Pueblo Nations: Eight Centuries of Pueblo Indian History*. Clear Light Publications, Santa Fe.

Sauer, Carl O.
1952 *Agricultural Origins and Dispersals*. Bowman Memorial Lectures, Series 2. The American Geographical Society, New York.

Sauer, Jonathon D.
1950 Amaranths as Dye Plants Among Pueblo Indians. *Southwestern Journal of Anthropology* 6(4):412–415.

Schaafsma, Polly
1992   Rock Art in New Mexico. Museum of New Mexico Press, Santa Fe.

Schoenwetter, James and Alfred E. Dittert, Jr.
1968   An Ecological Interpretation of Anasazi Settlement Patterns. In *Anthropological Archeology in the Americas*, ed. by Betty J. Meggers. The Anthropological Society of Washington, Washington, D. C.

Schroeder, Albert H. and Dan S. Matson
1965   *A Colony on the Move: Gaspar Castaño de Sosa's Journal, 1590–1591.* School of American Research Press, Santa Fe.

Simmons, Alan H.
1986   New Evidence for the Early Use of Cultigens in the American Southwest. *American Antiquity* 51:73–88.

Simmons, Marc
1987   The Spaniards of San Gabriel. In *When Cultures Meet*. Papers from the October 20, 1984, conference held at San Juan Pueblo, New Mexico. Sunstone Press, Santa Fe.

Smiley, T. L., S. A. Stubbs and B. Bannister
1953   A Foundation for the Dating of Some Late Archaeological Sites in the Rio Grande Area, New Mexico. *University of Arizona Bulletin*, Vol. 24, No. 3; Laboratory of *Tree-ring Research Bulletin* No. 6, Tucson.

Standley, Paul C.
1912   Some Useful Native Plants of New Mexico. *Smithsonian Institution Report, 1911.*

Steen, Charles R.
1982   *Pajarito Plateau Archeological Survey and Excavations, II.* Los Alamos National Laboratory, Los Alamos.

Stevenson, Matilda Coxe
1912   *Ethnobotany of San Ildefonso and Santa Clara Pueblos.* Unpublished ms., Ms. 4711, Archives of the Office of Anthropology, Smithsonian Institution, Washington, D.C.

1915   *Ethnobotany of the Zuni Indians.* Thirtieth Annual Report of the Bureau of American Ethnology, Washington, D.C.

Stiger, Mark A.
1977   *Anasazi Diet: The Coprolite Evidence.* Unpublished M.A. thesis, Univ. of Colorado, Boulder.

Stuart, David E.
1989   *The Magic of Bandelier.* Ancient City Press, Santa Fe.

Stuart, David E. and Robin E. Farwell
1983   Out of Phase: Late Pithouse Occupations in the Highlands of

New Mexico. In *High Altitude Adaptations in the Southwest*, ed. by Joseph C. Winter. Cultural Resources Management Report 2, USDA Forest Service, Southwestern Region, Albuquerque.

Stuart, David E. and Rory P. Gauthier
1981 *Prehistoric New Mexico: Background for Survey*. Historic Preservation Bureau, New Mexico Department of Finance and Administration, Santa Fe.

Swank, George R.
1932 *The Ethnobotany of the Acoma and Laguna Indians*. Unpublished M.A. thesis, Univ. of New Mexico, Albuquerque.

Tierney, Gail D.
1973 Plants and Man in Prehistoric Catron County. *El Palacio* 79(2):28–38.

1976 Of Pots and Plants. *El Palacio* 82(3):48–52.

1977 Plants for the Dyepot. *El Palacio* 83(3):28–35.

1979 Native Edible Plant Resources near Cochiti Reservoir, New Mexico. In *Archeological Investigations in Cochiti Reservoir, New Mexico, Vol. 4: Adaptive Change in the Northern Rio Grande Valley*, ed. by J. V. Biella and R. C. Chapman. Office of Contract Archeology, Univ. of New Mexico, Albuquerque.

1983 *Roadside Plants of Northern New Mexico*. The Lightning Tree, Santa Fe.

1987 The Archeology and History of Some New Mexico Weeds. In *Proceedings of the Southwestern Native Plant Symposium*, ed. by Lisa Johnston and Judith Philips. Native Plant Society of New Mexico, Albuquerque.

Toll, Mollie S.
1988 *Botanical Studies at an Extensive 13th Century Pithouse Village in the Southern Rio Grande Valley, New Mexico*. MNM Project No. 41.386. Castetter Laboratory for Ethnobotanical Studies, Technical Series No. 235. Univ. of New Mexico, Albuquerque.

Trierweiler, William Nicholas
1987 *The Marginal Cost of Production in Subsistence Economies: An Archeological Test*. Unpublished Ph.D. dissertation, Univ. of California, Los Angeles.

Turney, J. F.
1948 *An Analysis of Material Taken from a Section of Group M of the Cliffs, Frijoles Canyon, Bandelier National Monument, New Mexico*. Unpublished M.A. thesis, Adams State College, Alamosa, CO.

Underhill, Ruth
1979   Pueblo Crafts. The Filter Press, Palmer Lake, CO.

Wendorf, Fred and Erik. K. Reed
1955   An Alternative Reconstruction of Northern Rio Grande Prehistory. El Palacio 62(5-6):131–173.

Wetterstrom, Wilma
1986   Food, Diet, and Population at Prehistoric Arroyo Hondo Pueblo, New Mexico. School of American Research Press, Santa Fe.

Wheat, Joe Ben
1979   Rio Grande, Pueblo, and Navajo Weavers: Cross-cultural Influence. In Spanish Textile Tradition of New Mexico and Colorado, ed. by Nora Fisher. Museum of New Mexico Press, Santa Fe.

White, Leslie A.
1942   The Pueblo of Santa Ana, New Mexico. Memoirs of the American Anthropological Association, 60.

1945   Notes on the Ethnobotany of the Keres. Papers of the Michigan Academy of Science, Arts, and Letters 30:557-568.

Wills, W. H.
1988   Early Prehistoric Agriculture in the American Southwest. School of American Research Press, Santa Fe.

Winship, George Parker
1922   The Journey of Coronado, 1540-1542. Allerton Book Co., New York.

Winter, Joseph C.
1974   Aboriginal Agriculture in the Southwest and Great Basin. Unpublished Ph.D. dissertation, Anthropology Department, Univ. of Utah, Salt Lake City.

1976   The processes of Farming Diffusion in the Southwest and Great Basin. American Antiquity 41(4):421-429.

Wislizenus, Frederick A.
1848   A Tour to Northern Mexico, 1846-1847. 30th Congress, 1st Session, Misc. No. 26. Facsimile published by The Rio Grande Press, Inc., Glorieta, NM.

Yarnell, Richard
1958   Implications of Pueblo Indian Ruins as Plant Habitats. Unpublished M.S. thesis, Univ. of New Mexico, Albuquerque.

1959   Prehistoric Pueblo Use of Datura. El Palacio 66:176–178.

PHOTOGRAPHY CREDITS

All botanical line illustrations by Gail D. Tierney. All photographs by William W. Dunmire except as noted below.

Pg. 19, photo by Blair Clark, Museum of Indian Arts and Culture/Laboratory of Anthropology Collections, cat. nos. 15985, 16008. Pg. 20, Museum of New Mexico Photo Archives, neg. no. 66712. Pg. 32, photo by David Noble, courtesy School of American Research. Pg. 42, photo by Blair Clark, Museum of Indian Arts and Culture/Laboratory of Anthropology Collections, cat. nos. 43318/11, 41830/11, 43309/11. Pg. 53, photo by Jesse L. Nusbaum, Museum of New Mexico Photo Archives, neg. No. 8742. Pg. 60, photo by Blair Clark, School of American Research Collections in the Museum of New Mexico, cat. no. 6464/11. Pg. 60, photo by Douglas Kahn, Museum of Indian Arts and Culture/Laboratory of Anthropology Collections, cat. no. 27184/11. Pg. 61, photo by Douglas Kahn, Museum of Indian Arts and Culture/Laboratory of Anthropology Collections, cat. no. 49969/12. Pg. 64, photo by Douglas Kahn, Museum of Indian Arts and Culture/Laboratory of Anthropology Collections, cat. no. 16116 Pg. 65, Archives of the Laboratory of Anthropology/Museum of Indian Arts and Culture, photo no. 1116.58. Pg. 68–69, photo by Blair Clark, Museum of Indian Arts and Culture/Laboratory of Anthropology Collections, cat. nos. 18036/11, 18037/11. Pg. 70, courtesy National Park Service, Chaco Culture National Historical Park. Pg. 70, photo by Blair Clark, Museum of Indian Arts and Culture/Laboratory of Anthropology Collections, cat. no. 24777/12. Pg. 71, photo by Blair Clark, Museum of Indian Arts and Culture/Laboratory of Anthropology Collections, cat. no. 23131/12. Pg. 90, photo by Blair Clark. Pg. 243, photo by T. Harmon Parkhurst, Museum of New Mexico Photo Archives, neg. no. 5454.

INDEX

Abert, James W. 81, 235
Abies concolor (white fir), 102, **103**; bark/resin, 104; petroglyph of, 68; tanning, 63
Acequias (ditches), 43
Achillea lanulosa (yarrow), **221, 222**
Acoma Pueblo, 45, 55, 238; arrows, 137, 149; bows, 119; brushes, 63, 126; food plants, 144, 154, 174, 233; medicinal plants, 104, 107, 123, 137, 146, 170, 179, 197, 199, 201, 218, 220; smoking, 177; spindles, 137
Acorns, 16, 18, 21, 25, 39, 41, 114–16
Agave, ix, 82
Agriculture, 51, 80; development, 21, 22, 23, 24, 26; flood plain, 16, 23; impact on gathering, 56. See also Fields;Gardens
Alcoholic drinks, 133
Alder: dye from bark, 62
Alfilaria (Erodium cicutarium), 38
Allen, Patricia, 207
Allium cernuum (nodding onion), **163, 164**
Amaranthus spp., 19, 29, 171, **173, 174**, 175, 220; dye, 175; encouragement of, 53, 54, 175; food coloring, 62; genetics, 233; grain vs. vegetable, 173–74; harvesting, 50, 57; nutrition, 56, 175; pollen, 172
-cruentus, 175
-graecizans, 173
-hybridus, 173
-retroflexus, 173

Anasazi: diet, 34; origins, 23; sandals, 21, 60, 61, 126; term, 23–24
Anasazi Heritage Center, 253
Andropogon scoparius (little bluestem), **160, 161**
Angelica spp., 43
Antelope-sage (Eriogonum jamesii), 74, 167, 168
Anthropogenic material, 231
Apache plume (Fallugia paradoxa), 50, **134, 135**; arrows, 66, 134; brooms, 134–35; hair wash, 135
Apocynum spp. (dogbane): fibers, 61
Apricots, 41
Archaeology, 16, 55, 230
Archaic Period, 17–18, 23,77; foods, 19, 157, 159
Arrhenius, Olof, 75–-76
Arrows, 66, **68-69**; colors, 98; feathers, 98; glue, 98; tips, 7, 131, 137, 149, 159; wood for, 66, 116, 121, 131, 133, 134, 137, 139, 149, 159
Arroyo Hondo Pueblo, 4, 30–31, **32**, 33; corn racks, 110; occupation, 31, 34; plants, 33, 118, 154, 159
Artemisia spp. (sagebrush), 151–52; camphor, 152, 153
-filifolia (sand), **150**, **151**, 152
-frigida (fringed), **150**, 152
-tridentata (big), 74, **150**, 151, 152
Asclepias spp. (milkweed), 50, **196**, 197; fiber 61, 197

-asperula, **197**
-latifolia, 197
-subverticillata, 197
Aspen (Populus tremuloides), 111, 112, **113**; for drums, 67
Astragalus spp. (milkvetch), 74
Atriplex spp. (saltbush):
-argentea, 131
-canescens (fourwing), 62, **129, 130**, 131; arrows, 7, 66, 131; indicator plant, 7, 76, 78, 79, 129-30, 149
Axes/adzes, 42; metal, 97
Aztec Ruins National Monument, 252

Baldness prevention, 107
Bandelier, Adolph, 236
Bandelier National Monument, 4, 87,117, 143, 148, 223; Capulin Canyon, 118; Ceremonial Cave, **47**; ecology, 9-11, 17; Falls Trail, 5, 87, 122, 134, 151, 167; Frey Trail, 5, 87, 125, 136, 167, 219; petroglyph, 68; settlement, 26, 29; species count, 9; Talus House, 204; Upper Crossing Trail, 5, 87, 102, 115, 120, 125, 136, 164, 196, 202. See also Frijoles Canyon; Ruins Trail; Tsankawi Trail/ Ruins
Bark, 43; dye, 62, 137; medicinal uses, 104, 107, 110, 119
Barley, 38, 43
Baskets, 107; beargrass, 127–28; dyes, 149, 184; sumac, 139; willow, 21, 59, **61**, 108, 109-10; yucca, 59, **60**, 110, 111, 126

281

Beans, 3, 31, 34, 41, 44;
  cultivation, 27, 51, 81,
  235; nutrition, 56,
  115–16; origins,
  21–22, 38
Beargrass (*Nolina microcarpa*), **127**, 128;
  weaving with, 59, 61
Bee-balm (*Monarda menthaefolia*), 43, 50, **202**,
  203
Beeplant, Rocky
  Mountain (*Cleome serrulata*), 80, **182**, **183**; as
  food, 29, 50, 57, 175,
  182–83, 184; pigment
  (guaco), 29, 63, **64**,
  183–84, 190
Beets, 41
Berries, 34, 43, 106. *See
  also* Juniper
Birthing herbs, 107, 146
Bluegrass (*Poa* spp.), 76
Bluestem, little
  (*Andropogon scoparius*),
  160, **161**
Blue trumpets (*Ipomopsis
  longiflora*), 50, **198**, **199**
Boats, 112
Bohrer, Vorsila I., 238
Bosque (forest), 242
Bouteloua (grama grasses),
  81, 235
  -*curtipendula* (sideoats), **160**, **161**, **162**
  -*eriopoda* (black), 81
Bows, 66, 107, 116, 119,
  121, 133
Bromus tectorum
  (cheatgrass), 80
Brooms, 50, 67, 134–35,
  161
Brushes, 63, 126, 161
Buckwheat, wild
  (*Eriogonum* spp.), 57,
  59, **167**, 168
  -antelope-sage (*jamesii*), 74, 167, **168**
Building materials:
  insulation, 159; lath,
  154; *jacal*, 42; thatching, 110; timber, 42,
  64-66; 112; vigas
  (beams), 64, 100, 104

Cabbage, 41
Cabeza de Baca, 233–34

Cactus, 50, 189–91, 235;
  claret cup, 191;
  nutrition, 56; *tunas*
  (fruits), 33, **34**, 35,
  62, 190–91
  -cane cholla (*Opuntia
  imbricata*), 23, 33, 76,
  77, **78**, **140**, **141**,
  142, 232, 243;
  -hedgehog (*Echinocereus*
  spp.), **189**, 191
  -pincushion
  (*Coryphantha* spp.),
  189
  -pincushion
  (*Mammillaria* spp.), 189
  -prickly pear (*Opuntia
  phaeacantha*), 23, 33,
  34, 35, 87, **189**, **190**,
  191, 234
Cajete, Greg, 47, 48
Camas, death (*Zygodenus*
  spp.), 164
Camphor, 151, 152
Canaigre (*Rumex
  hymenosepalus*), 63, 169,
  170
Cantaloupes, 39, 41
Capsicum annuum, (chile),
  41, 44, 51, 54
Carrots, 41
Castañada chronicles, 39
Castaño de Sosa, Gaspar,
  38, 234
Castetter, Edward F., 238
Castilleja integra (paintbrush), **212**, **213**
Cattail (*Typha* spp.)
  -*angustifolia* (narrow-
  leafed), 154
  -*latifolia* (broad-
  leaved), 61, **153**, **154**
"Cedar," 41
Celery, wild (*Cymopterus
  fendleri*), 18, 57, **192**,
  193, 194
Cercocarpus montanus
  (mountain-
  mahogany), 62, 66,
  133, **136**, **137**
Ceremonial plants, ix, 58,
  103, 147, 153, 205,
  226
Chaco Canyon, 25, **70**,
  77; Pueblo Bonito, **65**,
  66; utilized plants,
  168, 174, 187, 200

Chaco Culture National
  Historical Park, 252
Chamisa, 148. *See also*
  Rabbitbrush
Chavez, David, 53, 175,
  211
Cheatgrass (*Bromus
  tectorum*), 80
Cheese-making, 209
Chenopodium spp. (goosefoot), 29, **171**;
  food, 19, 50, 172,
  175; nutrition, 56
  -*album* (lamb's quarters),
  38, 171
  -*leptophyllum*, **172**
Cherry, 41, 235
Chile (*Capsicum annuum*),
  41, 51, 54; introduction, 44
Chokecherry (*Prunus
  virginiana*), 34, **117**,
  118, 119, 125; bows,
  66, 119
Cholesterol, 56
Cholla: fences, 78, 142;
  -cane (*Opuntia
  imbricata*), 23, 33,
  140,**141**, 142,
  232, 243; as indicator
  plant, 76, 77, 78
Chrysothamnus nauseosus
  (rabbitbrush), 139,
  148, 149; dye, 62,
  148–49; galls, 149
Cilantro (coriander
  greens), 39
Clay: alkaloids and, 211
Cleome serrulata (Rocky
  Mountain beeplant),
  80, **182**, **183**; as food,
  29, 50, 57, 175,
  182–83, 184; pigment
  (guaco), 29, 63, **64**,
  183-84, 190
Climate, 14, 16-17, 35,
  241-42
Clothing, 3, 42, 126
Clover, white, 74
Cochiti Dam/Reservoir,
  220, 246; species
  count, 10
Cochiti Pueblo, 4, 236;
  acorns, 115; cordage,
  135; cradleboard, **70**;
  drums, 67, **71**, 112;
  dyes, 137; food plants,

282 INDEX

194; insect repellents, 215; medicinal plants, 107, 135, 146, 149, 152, 209, 220, 222, 224; popcorn, 233; soap/shampoo, 126, 216
Cocklebur (*Xanthium strumarium*), **217**, **218**, 226
Collecting (plant), 18, 59, 228
Common reed (*Phragmites communis*), **158**, **159**; arrows, 66, **68–69**, 159; in building, 65, 66, 159
Confidentiality (sacred plants and), ix, 8, 49
Conservation: Native Seeds/SEARCH, 54
Conservation ethic, 50–51
Cook, Sarah Louise, 238
Coprolite (feces) studies, 29, 33, 55, 117, 123, 161, 168, 174, 178, 184, 226
Cordage, 62, **70**, 109, 125–26, 135, 197
Coriander, 39, 53, 207; greens (*cilantro*), 39; introduction, 41; seeds (*coriandro, culantro*), 39
Coriandro (coriander seeds), 39 .
Corn, 3, 31, 41, 44, 81; cultivation, 27, **52**, 54, 144; depictions of, 68; development, 21–22, 23, 38, 233; heirloom, 233; importance, 27, 31, 34, 56; nutrition, 115, 130; popcorn, 233; *teosinte*, 21
Coronado, Francisco Vásquez de, 37, 39
Coronado State Monument, 80, 89; Kuaua ruins, 30, 89, 90, 103, 181, 200, **201**; vegetation, 112, 150, 155, 169, 181, 208
Corrales Bosque Preserve, 242
Cortés, Hernán, 40

Coryphantha spp. (pincushion cactus), **189**
Cotton (*Gossypium* sp.), 3, 62, 148; dyes for, 139, 148; growing, 28, 38; introduction of, 61, 38, 231; weaving, 35, 42
Cottonwood (*Populus* spp.); beams, 65, 66; digging sticks, 70; drums, 67, 71, 112; habitat, 74, 242;
-Fremont (*fremontii*), **111**, **112**, 113
-narrowleaf (*angustifolia*), **111**, 112
Cradleboard, 66, **70**, 100–01, 139; lashings, 126
Croton oil, 57, 185, 186
Croton texensis (doveweed), 57, **79**, 80, **185**, **186**
Cucurbitacins, 215
Cucurbita foetidissima (buffalo gourd), 80, **214**, **215**, 216; saponin in, 216
Culantro (coriander seeds), 39
Cultivation, plant: 22
Currant, wild (*Ribes inebrians*), 34, 66, **132**, **133**
Cushing, Frank Hamilton, 179, 236-37
Cymopterus spp.
-*bulbosus* (wafer parsnip), **192**, 193, 194
-*fendleri* (wild celery), 18, 57, **192**, **193**, 194
-*montanus* (wafer parsnip), **192**
-*purpureus* (wafer parsnip), **192**
-wild parsley, **192**
Cypress, summer (*Kochia scoparia*), 76, 80

Dalea formosa (feather dalea), 59
Darwin, Charles, 74-75, 78
Datura, sacred, 205
Datura meteloides (jimson-weed), 80, **204**, **205**
Dean, Glenna, 231
Decoration, plants in, 67–68
Deforestation, 43
Dendrochronology, 231–32
Deodorant, 139
Descurainia spp. (tansy mustard), 80; paint from, 63, 181
-*pinnata* (western tansy mustard), 77, **180**, **181**
"Diapers," 107
Digging sticks, 66, **70**, 107, 116, 137
Diversity, 9–11, 18
Dock (*Rumex* spp.), 58, 104, **169**, 170; dye, 18, 170; paint, 63; tanning, 63, 104, 123
-*canaigre* (*hymenosepalus*) 63, 169, **170**
-*crispus*, 170
Dogbane (*Apocynum* spp.) fibers, 61
Domesticated animals, 42
Domestication (plant), ix, 16, 22
Domingues-Escalante expedition, 234
Douglas-fir (*Pseudotsuga menziesii*), 67, 68, **102**, 103, 104; wood, 65, 66
Doveweed (*Croton texensis*), 57, **79**, 80, **185**, **186**
Dropseed (*Sporobolus* spp.), 19, 34, 77, **160**, 161
Drums, 67, **71**, 112
Dryland farming, 27, 29, 32
Dyes, 62, 170, 175, 190, 220, 224; food coloring, 62, 131; for leather, 62, 137, 149, 190, 213; for textiles, 62, 139, 148, 149
Echinocereus spp. (hedgehog cactus), **189**, 191
-*triglochidiatus* (claret cup cactus), 191
Ecozones, 91–93

INDEX **283**

Emory, W.H., 235
Engelmann, George, 236
Entrada, 35, 37, 40, 41
Ephedra spp. (joint-fir), 104, **122**, **123**
  -*nevadensis* (Morman tea), 63, 122
Ephedrine, 123
Eriogonum spp. (wild buckwheat), 57, 59, 167, 168
  -*jamesii* (antelope-sage), 74, 167, 168
Erodium cicutarium (alfilaria), 38
Ethnobotany, 227, 228–31; paleo-, 231, 238
Ethnology, 229

Fallugia paradoxa (Apache plume), 50, **134**, **135**; arrows, 66, 134; brooms, 67, 134–35; hair wash, 135
Feathergrass, New Mexico (*Stipa neomexicana*), 19, **20**, **162**
Fences: cholla, 78, 142; colonists', 42
Fiber: cordage, 62, **70**, 109, 125–26, 135, 197; prehistoric, 3, 35, 59, **60**, **61**, 62
Fields, 27, 81;indications of old, 80, 82, 149, 181, 182, 201, 225. *See also* Gardens
Fir, white (*Abies concolor*), 41, **102**, **103**; bark/resin, 104; in petroglyph, 68; tanning, 63
Fire, ix, 139, 244–45; La Mesa Fire, 120, 244, **245**
Firewood, 42
Flood plain agriculture, 16, 23
Flutes: sunflower, 226
Folk heroes, 229
Food plants, 55–57; species count, 56
Foraging, 18
Ford, Richard I., 21, 165, 238; San Juan studies, 50

Four o'clock, wild (*Mirabilis multiflora*), 55, **176**, **177**
Freire-Marreco, Barbara, 237
Fresnel Shelter, 176
Frijoles Canyon, 29, 30, **79**, 80, **88**, 108, 118, 126, 236; Long House, 7, 30, **78**, 204; Saltbush Ruin, 96, 130; Ski Trail, 102, 221;Tyuonyi, 30, **31**, 174, 182; vegetation, 7, 100, 112, 115, 120, 122, 125, 132, 167, 171, 173, 178, 181, 185, 206, 221
Fruits, 41, 44, 235
Fuel, 107, 113, 242, 246
Fumigation plants, 107, 152

Gambel, William, 234
Gardening, 54; transplanting, ix, 50, 203. *See also* Fields; Gardens
Gardens, 41, 80–82, 177, 225, 235; grid, 27–28, 30, 81–82; pebble mulch, 81, 231; stone-outlined, 42, 81, 231; waffle, **52**, **53**, 175
Gathering (wild plant), 15, 18, 33, 41; agriculture and, 56; seasonal, 56, 57
Geranium, wild: tannin, 58
Ghost Ranch Living Museum, 251
Globe-mallow (*Sphaeralcea* spp.), 53, **187**
  -*coccinea* (scarlet), **188**
Glue, 66, 97-98, 187
Gooseberry (*Ribes inerme*), 34, **132**, **133**, 139
Goosefoot (*Chenopodium* spp.), 29, **171**; food, 19, 50, 172, 175; nutrition, 56
  -lamb's quarters (*album*), 38, 171
  -*leptophyllum*, **172**
Gossypium. (cotton), 3, 62, 148; dyes for, 139,

148; growing, 28, 38; introduction of, 38, 61, 231; weaving, 35, 42
Gourds, 27, 54, 234 -bottle (*Lagenaria siceraria*) 38, 54, 215
  -buffalo (*Cucurbita foetidissima*), 80, **214**, **215**, 216; saponin in, 216
Grafting, 41
Grama grass (*Bouteloua* spp.), 235
  -black (*eriopoda*), 81
  -side-oats (*curtipendula*) **160**, **161**, **162**
Grapes, 41
Grasses, 23, 34, 160–61
Grasslands, 17, 18
Greasewood (*Sarcobatus vermiculatus*), 77
Greenthread, 223
Grid gardens, 27–28, 30, 81–82
Grindelia aphanactis (gumweed), **219**, **220**
Grinding implements, 16, 18; mano and metates, **19**
Groundcherry (*Physalis* spp.), 23, 50, 53, 54, 57, 80, 143, **206**, 207
  -*hederaefolia*, **207**
  -*philadelphica* (husk tomatillo), ix
Gum, chewing, 197
Gumweed (*Grindelia aphanactis*), **219**, **220**
Gutierrezia sarothrae (broom snakewood), 77, **145**, **146**, 147, 243

Habitats (plant), 9-11, 14, 241; access to, 59; changes, 17, 80, 228, 242-47; overgrazing, 156, 181, 242, **243**, 244; preservation, 59; wood cutting, 97, 245
Harrington, John P., 237
Harshberger, John W., 227
Helianthus annuus (sunflower), 50, 53, 57, 179, **225**, **226**

Heron's bill (*alfilaria*), 38
Historical data, 233–35
Hoes, 42
Homesteaders, 80–81
Honeybee: introduced, 74
Hopi, 131, 170, 224
Horse-nettle (*Solanum elaeagnifolium*), 80, **208**, **209**, 210
Horses, 37, 38
Hughes, Phyllis, 226
Hunters and gatherers, 3, 16, 18

Implements, 3, 66–67, 110, 116, 131, 136–37; digging sticks, 41, 66, 116; fire sticks,110;grinding, 16, 18, 19; hoes, 41, 42; iron, **42**
Indian Pueblo Cultural Center, 250
Indian ricegrass (*Oryzopsis hymenoides*), 18, 19, 34, 51, 77, **155**, **156**, 157 172, 179; habitat reduction, 243
Indian tea (*Thelesperma* spp.), 52, 53, 55, 57, 62
-*filifolium*, 80
-*megapotamicum*, 50, 123, **223**, **224**
Indicator plants, 73–80, 149, 181, 200–01
Insect repellents, 215, 216
Insulation, 159
Introduced plants, 38, 41, 44, 171
Invasive plants, 29, 38
Ipomopsis
-*longiflora* (blue trumpets), 50, **198**, **199**
-*multiflora* (many-flowered), 199
Iron tools, **42**
Irrigation, 27, 32, 42, 43
Isleta Pueblo, 154, 238 arrows, 131, 149; boats, 112; bows, 119; cactus, 191; food plants, 104, 113, 128, 144; heirloom corn, 233; medicinal plants, 123, 146, 149, 152, 164, 186, 197, 209, 216

Jacal (upright posts), 42
Jemez Cave, 4, 139, 152, 159; corn, 20, 21
Jemez Pueblo, 4, 58, **90**, 159, 213, 238; baskets, 59, 139; bows, 121; drums, 112; food plants, 113, 121; gardening, 81; medicinal plants, 146, 149, 152, 186, 218, 226; paint, 97; shampoo, 126
Jemez State Monument, 80, 89; vegetation, 106, 132, 134, 144, 148, 155, 187, 196, 207, 208, 214, 219, 220
Jimsonweed (*Datura meteloides*), 80, **204**, **205**
Joint-fir (*Ephedra* spp.), 104, **122**, **123**; "Mormon tea," 122
Jones, Volney H., 144, 191, 238
Juniper (*Juniperus* spp.), 59, 107, 229, 245; berries, 105, 106, 107, 234; medicinal uses, 107, 146, 229; wood, 64, 66, 107
-alligator (*deppeana*), **105**, 106
-one-seed (*monosperma*), 11, 64, **105**, **106**, 232, 243
-Rocky Mountain (*scopulorum*), **105**, 106
Juniper-grassland ecozone, 92, 93
Juniper-mistletoe (*Phoradendron juniperinum*), 107

Kachinas, 113
Kent, Kate Peck, 61
Kochia scoparia (summer cypress), 76, 80
Kuaua ruins, 30, 89, **90**, 103; vegetation, 181, 200, **201**

Lagenaria siceraria (bottle gourd), 38, 54, 215
Laguna Pueblo, 45, 119, 137, 149, 238; food plants, 119, 144, 154, 174; medicinal plants, 107, 123, 146, 170, 179, 186, 197, 201, 218, 220; smoking, 174
Lamb's quarters (*Chenopodium album*), 38, 171
Land issues, 47-49, 51, 59
Lange, Charles, 137
Latillas (wood lath), 66
Lauriano, Felipe, 98, 137, 209, 216
Lemonade-bush, 138, 234
Lentils, 41
Lettuce, 41
Lichens: paint from, 63
Ligusticum porteri (osha), 43
Lithospermum spp., 58
Liverwort (*Marchantia* spp.), 76
Locust, New Mexico (*Robinia neomexicana*), 66, **120**, **121**
Loncasion, Lorrain, 52, 53
Lovage (*Angelica* spp.), 43
Lycium pallidum (wolfberry, tomatillo), 76, **143**, **144**
Lye, 130

Maize. *See* corn.
Mammillaria spp. (pincushion cactus), **189**, 190, 191
Marchantia spp. (liverwort), 76
Martin, Bill, 67
Martin, Paul S., 15
Maxwell Museum of Anthropolgy, 250
Medicinal uses, 57-59; aches, 146, 177, 188, 213; antiseptics, 98, 170, 179, 208, 218, 229; arthritis, 194; bites, 107, 131, 146, 209, 226; burns, 170, 222; colds/chills/

INDEX **285**

coughs, 98, 107,
110, 119, 123, 146,
149, 152, 164, 197,
222, 224; constipation, 107, 137, 186,
209,216;cuts/ wounds/
sores, 5, 8, 98, 104,
119, 144, 168, 183,
188, 202, 218, 220,
226; diarrhea, 107,
123, 179, 218; earaches, 104, 107, 168,
186; emetics, 107,
146, 167, 199; for
eyes, 55, 146, 168,
202; fever, 146, 149,
197, 198, 202, 222;
headache, 186, 198,
199, 202; heart, 197,
202; indigestion, 55;
pain, 110, 113; pneumonia, 128, 164;
rheumatism, 55, 104,
107, 128, 146, 186;
Rocky Mountain
spotted fever, 152;
skin problems, 58,
113, 123, 131, 170,
186, 201, 205; stomach problems, 57,
107, 135, 149, 152,
167, 176, 183, 186,
197, 202, 209, 220,
222, 224; toothache,
110, 149, 209, 220;
urinary, 100, 113, 123,
146, 218;
vomiting/nausea, 58,
152, 218
Melons, 39–40, 54, 233
Mentzelia spp., (stickleaf)
39
Mesa Verde National
Park, 253
Mesquite, 74, 234
Metal, presence of, 42
Milkvetch (*Astragalus*
spp.), 74
Milkweed (*Asclepias*
spp.), 50, **196**, 197;
fiber 61, 197
-*asperula*, **197**
-*latifolia*, 197
-*subverticillata*, 197
Mints, 43, 57
Mirabilis multiflora (wild
four-o'clock), 55, **176**,

177
Mixed conifer ecozone,
92, **93**, 102
Monarda menthaefolia (beebalm), 43, 50, **202**,
203
Monoculture, 51
Moore, Michael, 116
Morman tea *See Ephedra*
spp.
Morning glory, 54
Mountain-mahogany
(*Cercocarpus montanus*),
62, 66, 133, **136**, **137**
Museum of Indian Arts
and Culture, 250
Mustard, 29, 180-81

Nabhan, Gary Paul, 54,
238
Nambé Pueblo, 55
Names (plant), 55, 86,
164
Native Seeds/SEARCH, 54,
233
Navajo, 223; dyes, 224
"Navajo sage," 152
Needlegrass (*Stipa* spp.),
160, 161-62, 243
-New Mexico feathergrass
(*neomexicana*), 19, **20**,
162
Nolina microcarpa (beargrass), **127**, **128**;
baskets, 127-28;
saponin in, 128; weaving, 59, 61
Nutritional values, 56–57.
See also Protein sources
Nuts, 56. See also Piñon
nuts

Oak, 39, 66, 123, 133,
159
-Gambel (*Quercus gambelii*), **114**, **115**, 116
234
Ojo Caliente: gardens, 81
Oñate, Juan de, 40
Onion, 41, 44, 53, 57
-nodding (*Allium cernuum*), **163**, **164**
Opuntia spp. (cactus):
tunas (fruits), 33, **34**,
35, 62, 190-91
-*imbricata* (cane
cholla), 23, 33, 76, 77,

78, **140**, **141**, 142,
232, 243
-*phaeacantha* (prickly
pear), 23, 33, 34, 35,
87, **189**, **190**, 191,
234
Orchards, 41, 42, 43, 235
Oregano, mountain, 202
Oregano de la sierra, 43
Oryzopsis hymenoides
(Indian ricegrass), 18,
19, 34, 51, 77, **155**,
156, 172, 179;
cultivation, 157;
habitat reduction, 243
Osha (*Ligusticum poteri*),
43
Otowi Ruins, 4, 30
Oyenque, Ramos, 54,
175, 226

Paa-ko Pueblo, 30
Paint, 62-63, 97, 181, 213
guaco, 184; house
paint, 187; Rocky
Mountain beeplant,
29, 63, 64, 183, 184
Paintbrush (*Castilleja
integra*), **212**, **213**
Pajarito Plateau, 1, 4,
17, 92; settlement, 17,
24–26, 29–30
Paleoethnobotany, 238
Palynology (pollen
analysis), 232
Parsley
-mountain
(*Pseudocymopterus
montanus*), **192**, **194**,
195
-wild (*Cymopterus*
spp.), 192, 193
Parsnip, wafer (*Cymopterus*
spp.), **192**, 193, 194
-*bulbosus*, **192**
-*montanus*, **192**
-*purpureus*, **192**
Peaches, 41
Pears, 41
Peas, 41
Pecos National Historical
Park, 251
Pecos Pueblo, 67, 80,
181, 187
Petroglyph National
Monument, 20, 87;
Boca Negra Canyon,
68, 87, **88**, **125**; cel-

ery, 193; dock, 169; doveweed, 185; dropseed, 161; globemallow, 187;horsenettle, 208;Indian ricegrass, 155;Indian tea, 223; joint-fir, 122; Macaw Trail, 68; needlegrass, 161;plant depictions in, 68, **71**, 125; sagebrush, 150, 151; wolfberry, 144; yucca, 125
Petroglyphs, 68, **71**, 125
Phacelia spp. (scorpionweed), **200**, 201
  -*corrugata*, **200**, **201**
Phoradendron juniperinum (juniper mistletoe), 107
Phragmites communis (common reed), 65, 66, **158**, **159**; arrows, 66, **68–69**, 159
Physalis spp. (groundcherry), 23, 50, 53, 54, 57, 80, 143, **206**, 207
  -*hederaefolia*, **207**
  -*philadelphica*, (husk tomatillo), ix
Picuris Pueblo, 166, 196, 197; medicinal plants, 104, 152, 168, 188, 220, 222
Pigweed, 38, 173
Piki bread, 131, 175
Pino, Peter, 28, 51
Piñon-juniper ecozone, 11, 14, 17, 92, **93**, 95, 114, 232, 244; changes, 246
Piñon nuts, 18, 21, 23, 25, 56, 96; gathering, 57, 96, 97; nutrition, 97; processing, 16, 96, 97, 234; yields, 18, 33, 96
Piñon pine (*Pinus edulis*), 11, 41, 59, **95**, **96**, 98, 235; dye, 62, 139; fuel, 246; gum, 62, 66, 97, 98, 139; habitat, 245; paint, 63, 97; pitch, 63, 66, 97-98, 159; wood, 64, 66, 59
Pinus edulis. See Piñon pine

Pinus ponderosa. See Ponderosa pine
Pioneering plants, 50
Plant breeding, 228
Plowshares, 42
Plums, 41, 235
Poa spp. (bluegrass), 76
Pollen analysis, 16, 55, 141, 172, 231, 232
Ponderosa pine (*Pinus ponderosa*), 92, **99**, **100**; bark, 99, 100; cradleboard, **70**; wood, 65, 66, 100, 104
Ponderosa pine ecozone, 11, 17, 92, **93**, **114**; fire in, 244
Populus spp. (cottonwood), 11, 74; beams, 65, 66; digging sticks, **70**; drums, 67, **71**; 112; habitat, 74, 242
  -*angustifolia* (narrowleaf cottonwood), **111**, 112
  -*fremontii* (Fremont cottonwood), 111, 112, 113
Populus tremuloides (aspen), 67, **111**, 112, **113**
Portulaca spp. (purslane), 29, 50, 53, 54, 56
  -*oleracea*, 38, **178**, **179**
  -*retusa* (notchleaf), 38, 179
Potato, wild (*Solanum* spp.), 53, 208, 235
  -*fendleri*, **210**
  -*jamesii*, 80, **210**, **211**
Pot Creek Cultural Center, 252
Potter, Loren, 77
Pottery: firing fuel, 113. *See also* Paint
Powell, John Wesley, 236
Prayer sticks, 67, 103, 110, 137
Prickly pear (*Opuntia phaeacantha*), 23, 33, 87, **189**, **190**, 234; *tunas* (fruits), 33, **34**, 35, 62, 190–91
Protein sources, 19, 56, 97, 115, 157, 172
Prunes, 41

Prunus virginiana (chokecherry), 34, **117**, **118**, 119, 125; bows, 66, 119
Pseudocymopterus montanus (mountain parsley), **192**, **194**, 195
Pseudotsuga menziesii (Douglas-fir), 67, 68, **102**, **103**, 104; wood, 65, 66
Puccoon (*Lithospermum* spp): tannin in, 58
Pueblo Indians, 45–46; settlement, 2, 4, 10, 11; world view, 8, 46–49, 50
Pueblo Province:definition, 5; species count in, 6, 85
Purslane (*Portulaca* spp.), 29, 50, 53, 54, 56
  -*oleracea*, 38, **178**, **179**
  -*retusa* (notchleaf), 38, 179
Puye Ruins, 4, 30, 138, 204

Quelites (greens), 38
Quercus gambelii (Gambel oak), **114**, **115**, 116, 234

Rabbitbrush (*Chrysothamnus nauseosus*), 139, **148**, **149**; dye, 62, 148–49; galls, 149
Radishes, 41
Rattles, 128, 215
Reed, 65, 131
  -*Phragmites communis* (common), 65, 66, **68–69**, **158**, **159**
Rhubarb, wild, 169, 170
Rhus trilobata (threeleaf sumac), 21, 62, **70**, 82, **138**, **139**, 235; arrows, 66, 133, 139; baskets, 59, 139; lemonade-bush, 138, 234
Ribes inebrians (wild currant), **132**, **133**; arrows, 66, 133
Ribes inerme (gooseberry), 34, **132**, 133, 139

INDEX **287**

Ricegrass, Indian
(*Oryzopsis hymenoides*),
19, 34, 51, 77, **155**,
156, 172, 179;
cultivation, 157;
habitat reduction, 243
Rio Grande Valley, 17,
39; diversity, 9-11;
settlement, 4, 10, 11,
13, 23, 25, 34–35, 40
Riparian ecozone, **94**,
117, 242
Robbins, Wilfred William,
237
Robinia neomexicana (New
Mexico locust), 66,
120, 121
Rocky Mountain beeplant
(*Cleome serrulata*), 80,
182, 183; as food, 29,
50, 57, 175, 182-83,
184; pigment (guaco),
29, 63, 64, 183–84,
190
Roofing, 104, 110, 154;
vigas (beams), 64, 100,
104
Ruins Trail (Bandelier),
5, 7, 87; Apache
plume, 134; bee-balm,
202; beeplant, 182;
bluestem, 161; blue
trumpets, 198; four-
o'clock, 176;
globe-mallow, 187;
gumweed, 219; horse-
nettle, 208; jimson-
weed, 204; potato,
210
Rumex spp. (dock), 58,
104, 169, 170; dye,
18, 170; paint, 63;
tanning, 63, 104, 123
-*crispus*, 170
-*hymenosepalus*
(canaigre), 63, 169,
170

Sacred plants:
confidentiality and, ix,
8, 49
Sagebrush (*Artemisia* spp.)
-big (*A. tridentata*), 74,
150, 151, 152
-fringed (*A. frigida*),
150, 151, 152

-sand (*A. filifolia*), **150**,
151, 152
Salicin, 110
Salinas Pueblo Missions
National Monument,
251
Salix exigua (coyote willow), **108**, **109**, 110
Salmon Ruin Museum and
Heritage Park, 252
Salsola kali (Russian
thistle), 80
Saltbush (*Atriplex* spp.)
-*argentea*, 131
-*canescens* (fourwing),
62, **129**, **130**, 131;
arrows, 7, 66, 131;
indicator plant, 7, 76,
78, 79, 129–30, 149
Sandals, 21, **60**, 61, 126
Sandia Pueblo, 135, 137,
147, 216; food plants,
98, 144, 209; gum,
197; medicinal plants,
107, 149
San Felipe Pueblo: food
plants, 149, 154, 175
San Gabriel, 40
San Ildefonso Pueblo, 4,
55, 107, 119, 135,
226; baskets, **61**; food
plants, 113, 125, 175;
medicinal plants, 113,
144, 146, 152, 168,
183, 201; paint, **64**
San Juan Pueblo, 4, 40,
50, 55, 226; arrows,
137, 226; bows, 119,
121; food plants, 97,
113, 144, 175, 183,
194; Ford's study, 50,
51; gardens, 54; medi-
cinal plants, 100, 110,
146, 152, 164, 186,
197, 198,202, 220,
222; roofing, 104;
seed necklaces, 107
Santa Ana Pueblo, 146,
187
Santa Clara Pueblo, 4, 55,
115, 184; medicinal
plants, 107, 123, 144,
146, 149, 176, 188,
201, 202
Santa Fe Trail, 234
Santo Domingo Pueblo,
187, 216, 233; medici-

nal plants, 146, 218,
224
Saponin, 57, 126, 128,
131, 167, 216
Sarcobatus vermiculatus
(greasewood), 77
School of American
Research (SAR), 30
Scorpionweed (*Phacelia*
spp.), **200**, 201
-*corrugata*, 200, **201**
Seasonality, 17, 22, 34;
harvesting, 18, 21
Seed preservation, 54
Sheep sorrel, 169
Smilacina spp. (false
solomon's seal), **165**
-*racemosa*, **166**
Smith, Stan, 147
Smoking (plants for), 55,
159, 177
Snakeweed, broom
(*Gutierrezia sarothrae*),
77, **145**, **146**, 147,
243
Snuff, 209
Soap/shampoo, 51, 126,
131, 216
Solanum spp. (wild pota-
to), 53, 208, 235
-*fendleri*, **210** *jamesii*,
80, **210**, **211**
Solanum elaeagnifolium
(horse-nettle), 80,
208, **209**, 210
Solomon's seal, false
(*Smilacina* spp.), **165**,
166
-*racemosa*, **166**
"Sore-eye poppy," 188
Spanish settlement, 4,
35, 37, 40; plants, 38,
39, 41, 44
Species counts, 6, 9, 56,
85
Sphaeralcea spp. (globe-
mallow), 53, **187**
-*coccinea* (scarlet), **188**
Spinach, wild, 29, 183
Sporobolus spp. (dropseed),
19, 34, **160**, 161
Spruce-fir ecozone, 17,
41, 92
Squash, 3, 54; cultivation,
27, 51; Hubbard, 41;
origins, 21–22, 23, 38;
prehistoric, 21, 31, 34,

41, 44
Standley, Paul Carpenter, 237
Starvation foods, 33, 100, 106, 142, 159
Steen, Charles, 78
Stevenson, James, 236
Stevenson, Matilda Coxe, 144, 237
Stickleaf (*Mentzelia* spp.), 39
Stipa spp. (needlegrass) **160**, 161-62
-*neomexicana* (New Mexico feathergrass), 19, **20**, **162**
Storage (of food), 16, 18, 19, 22, 41
Stuart, David, 26
Sumac, threeleaf (*Rhus trilobata*), 21, 62, **70**, 82, **138**, **139**, 235; arrows, 66, 133, 139; baskets, 59, 139; lemonade-bush, 138, 234
Sunflower (*Helianthus annuus*), 50, 53, 57, 179, 225, 226
Swank, George R., 238

Tallgrass prairie, 17
Tannin, 58, 115, 116, 123
Tanning, 62, 63–64, 76, 104, 123, 170
Tansy mustard (*Descurainia* spp.), 180–81; indicator plant, 77, 80, 181; paint, 63, 181
-*pinnata*, **180**, **181**
Taos Pueblo, 237; drums, 112; vegetation, 151, 196
Teosinte, 21
Tesuque Pueblo, 113, 119, 139, 194, 213
Thatching, 110
Thelesperma spp. (Indian tea), **52**, **53**, 55, 57, 62 -*filifolium*, 80 -*megapotamicum*, 50, 123, **223**, **224**
Thirst quencher, 123
Thistle, Russian (*Salsola kali*)
Tijeras Pueblo, 30
Tomatoes, 41

Tomatillo (*Lycium pallidum*) (wolfberry), 76, **143**, **144**
Tomatillo (*Physalis* spp.), ix, 143, 207, 233
Tools. *See* Implements
Toya, Patrick, 58, 59, 139
Trade, prehistoric, ix
Tsankawi Trail/Ruins (Bandelier), 5, 87; Apache plume, 134; bluestem, 161; blue trumpets, 198; buckwheat, 167; caves, 200; four-o'clock, 176; Indian tea, 223; mountain-mahogany, 136; petroglyph, 68; ruins, 7, 30, 181; sagebrush, 151; saltbush, 7; scorpionweed, 200; sumac, 138; tansy mustard, 181; yucca, 125
Typha spp. (cattail), 61, 153-54
-*angustifolia* (narrow-leafed), 154
-*latifolia* (broad-leafed), **153**, **154**

Underhill, Ruth, 63, 133

Verdolaga (purslane), 38
Vigas (roof beams), 64, 100, 104

Waffle gardens, **52**, **53**, 175
Waquie, Corina, 59
Water-catchment gardens, 81
Watermelons, 39, 41
Weaving: looms, 61; tools, 116, 137. *See also* Baskets; Fiber
Wheat, 38, 41, 44
White, Carl, 81
"White man's foot" (white clover), 74
Wild plants, 4, 51, 54, 56–57; number utilized, 56
Willow, 58, 108–10, 159; baskets, 21, 59, **61**, 109–10; habitat, 242 -coyote (*Salix exigua*), **108**, **109**, 110

Wills, W.H., 22
Wislizenus, Frederick A., 235
Wolfberry (*Lycium pallidum*), 76, **143**, 144
Wood. *See* Building materials
Wooton, E.O., 237
Xanthium strumarium (cocklebur), **217**, **218**, 226

Yarnell, Richard, 80
Yarrow (*Achillea lanulosa*), **221**, **222**
Yucca spp., 51, **124**, **125**, 126; baskets, 59, **60**, 110, 111, 126; brushes, 63, 126; cordage, 62, **70**, 109, 125-26; depictions of, 68, **71**, 125; quids, 125; sandals, 21, **60**, 61, 126; soap, 51, 126; weaving, 42, 61, 126, 154
-*baccata* (banana yucca), 33, **124**, **125**, 246
-*glauca* (narrowleaf yucca), 59, **124**, 125, 126; brushes, 63, 126

Zia Pueblo, 28, 51, 174, 187; grid gardens, 30
Zuni Pueblo, 45, 55, 154, 236–37; basketry, 149, 161, 197; cholla, 142; dyes, 149, 224; food coloring, 175; foods, ix, 39, 62, 119, 142, 144, 157, 174, 179, 190, 203, 207, 211, 217, 233;gardens, 51, **52**, **53**, 175, 203; medicinal plants, 7, 57, 107, 110, 146, 152, 168, 170, 177, 186, 199, 201, 208, 216, 218, 222, 226; ritual uses, 103, 110, 126, 137; smoking, 159; soap, 131; tomatillo, ix, 207
Zygodenus spp.(death camas), 164

INDEX **289**

ABOUT THE AUTHORS

GRADUATE OF the University of California, Berkeley, with degrees in wildlife management and zoology, Bill Dunmire served twenty-eight years in the National Park Service, mostly as a naturalist in a number of national parks, followed by several years as a biologist with The Nature Conservancy. A professional nature photographer and author of numerous booklets and articles relating to natural history, Bill lives with his wife, Vangie, in Placitas, New Mexico.

GAIL TIERNEY was raised among the valleys and mesa of New Mexico that are featured in this book. Trained in anthropology and botany at the University of New Mexico, she has published numerous reports, popular articles, and an earlier book (*Roadside Plants of Northern New Mexico*, with Phyllis Hughes) about botany in the state. Mrs. Tierney is the very proud mother of four grown New Mexicans and lives with her husband in Santa Fe.